Geospatial Semantics and the Semantic Web

SEMANTIC WEB AND BEYOND
Computing for Human Experience

Series Editors:

Ramesh Jain
University of California, Irvine
http://ngs.ics.uci.edu/

Amit Sheth
Wright State University
http://knoesis.wright.edu/amit/

As computing becomes ubiquitous and pervasive, computing is increasingly becoming an extension of human, modifying or enhancing human experience. Today's car reacts to human perception of danger with a series of computers participating in how to handle the vehicle for human command and environmental conditions. Proliferating sensors help with observations, decision making as well as sensory modifications. The emergent semantic web will lead to machine understanding of data and help exploit heterogeneous, multi-source digital media. Emerging applications in situation monitoring and entertainment applications are resulting in development of experiential environments.

SEMANTIC WEB AND BEYOND
Computing for Human Experience
addresses the following goals:

➢ brings together forward looking research and technology that will shape our world more intimately than ever before as computing becomes an extension of human experience;
➢ covers all aspects of computing that is very closely tied to human perception, understanding and experience;
➢ brings together computing that deal with semantics, perception and experience;
➢ serves as the platform for exchange of both practical technologies and far reaching research.

For further volumes:
http://www.springer.com/series/7056

Naveen Ashish · Amit P. Sheth
Editors

Geospatial Semantics and the Semantic Web

Foundations, Algorithms, and Applications

 Springer

Editors
Naveen Ashish
University of California Irvine
4308 Calit2 Building
Irvine CA 92697
USA
ashish@ics.uci.edu

Amit P. Sheth
Ohio Center of Excellence in Knowledge-
  enabled Computing (Kno.e.sis)
Wright State University
3640 Colonel Glenn Highway
45435-0001 Dayton Ohio
USA
amit.sheth@wright.edu

ISSN 1559-7474
ISBN 978-1-4614-2908-1   ISBN 978-1-4419-9446-2 (eBook)
DOI 10.1007/978-1-4419-9446-2
Springer New York Dordrecht Heidelberg London

Springer is part of Springer Science+Business Media (www.springer.com)

Preface

The availability of geographic and geo-spatial information and services, especially on the open Web, has become abundant in the last several years with the proliferation of online maps, geo-coding services, geospatial Web services and geospatially enabled applications. Concurrently, the need for geo-spatial reasoning has significantly increased in many everyday applications ranging from personal digital assistants, to Web search applications and local aware mobile services, to specialized systems in critical applications such as emergency response, medical triaging, and intelligence analysis to name a few. In response to the required "intelligent" information processing capabilities, the field of *Geospatial Semantics* has emerged as an exciting new discipline in the recent years. Broadly speaking geospatial semantics can be defined as the area that focuses on the *semantics* aspect in geographic and geo-spatial information processing i.e., where we can provide "meaning" to and intelligence in such information systems. This new area brings together researchers from many different disciplines such as geographic and geo-spatial information science, artificial intelligence – in particular the Semantic Web, and information systems. Alternate descriptions of what geospatial semantics is about can be stated as being the sub-area of geographic or geospatial information systems that deals with knowledge driven or intelligent processing techniques, or the particular domain application of semantics technologies that deal with the geographic and geospatial domain. Work in this area was initiated just a few years ago by visionary researchers who foresaw the need for expanding erstwhile individual disciplines such as GIS or the Semantic Web. Despite being a nascent field by age, we have seen a prolific amount of activity in all arenas, be it basic research, technical product development, community efforts such as developing standards, or the realization of real-world applications powered by such technologies.

Our primary goal in assembling this collection of work in geospatial semantics is to provide a first of a kind, cohesive collection of recent research in the theme of geospatial semantics. Additionally we have sought to present descriptions of fundamentally new information systems *applications* that have a potential for high impact and commercialization, and that become realizable with geospatial semantic technologies. The discipline of geospatial semantics has really emerged from a marriage between the erstwhile three separate areas of (1) Geographic information systems (GIS) or geo-spatial information processing, (2) Semantic Web technologies, and (3) Applications that are driving the demand for such capabilities, especially in the context of rapidly increasing use of location-aware mobile devices. We believe that the present is an appropriate stage to attempt to consolidate and formally define the new discipline of geospatial semantics. The activity in this area has expanded the horizons of the existing disciplines of GIS, the Semantic Web, as well as key applications. GIS techniques are now embellished with semantics smarts, the Semantic Web technologies have found a new "killer application" in the geo-spatial and GIS domains, and fundamentally new kinds of capabilities are now becoming realizable in key information systems applications.

This collection is mix of chapters on topics in the geospatial semantics area covering foundational aspects, infrastructure, as well as innovative applications. The initial chapters cover foundational aspects on semantic modeling and representation. These are followed by semantic infrastructure related chapters on issues such as effective query languages as well spatial cyber-infrastructure. The last three chapters are focused on applications of geospatial semantic technologies in key areas, namely earth observation systems, location based access control and major geo-informatics applications such as The National Map.

Chapter 1 presents an approach to representing and maintaining a *time series* of spatial ontologies, that is aimed at addressing the problem of retrieval of information with a geospatial context but at possibly different times. Place names and their geographical coverage evolve and change with time, and the time series capability at the ontology level is presented as the approach to achieving accurate information retrieval with such evolution.

Chapter 2 provides an approach to dealing with semantics of geoinformation in terms of *observable* properties. The thesis in the chapter is that observations are the principal source of geographic information and the semantic representation of such observations at the appropriate abstraction level is a key challenge that must be addressed.

Chapter 3 presents SPARQL-ST, an extension of the SPARQL query language, for handling *complex* spatio-temporal queries over semantic data.

Chapter 4 is concerned with geospatial semantic infrastructure, in particular considering spatial data infrastructures (SDI) as the basis for geospatial semantic interoperability. Overall this work is concerned with the development of a path towards realizing a spatial cyber-infrastructure.

Chapter 5 takes a key application area, that of earth observation systems (EOS) and provides an approach for incorporating semantic awareness in such systems. The approach is based on using ontologies to provide a semantic interpretation of the data collected by such earth observations systems in general.

Chapter 6 provides an approach to addressing access control in the context of location based applications. An access control system based on the role-based access control (RBAC) mechanism is presented that enforces *location* as well as context aware access control policies.

Finally Chap. 7 presents a description of the incorporation of semantics and semantic technologies in the important *National Map* effort. The chapter represents an important case study on the incorporation of semantics into a key geospatial information system namely The National Map.

Contents

Contributors

Jonathan Carter United States Geological Survey, Rolla, MO, USA, jjcarter@usgs.gov

Isabel F. Cruz ADVIS Lab, Department of Computer Science, University of Illinois at Chicago, Chicago, IL, USA, ifc@cs.uic.edu

Anusuriya Devaraju Institute for Geoinformatics, University of Münster, Münster, Germany, anusuriya.devaraju@uni-muenster.de

Rigel Gjomemo ADVIS Lab, Department of Computer Science, University of Illinois at Chicago, Chicago, IL, USA, rigelgjomemo@yahoo.com; rgjomemo@cs.uic.edu

Carlos Granell Institute of New Imaging Technologies, Universitat Jaume I, Castellon de la Plana, Spain, carlos@lsi.uji.es

Francis Harvey Department of Geography, University of Minnesota, Minneapolis, MN, USA, fharvey@umn.edu

Gobe Hobona Centre for Geospatial Science, University of Nottingham, Nottingham, UK, gobe.hobona@envitia.com

Eero Hyvönen Aalto University, Aalto, Finland, Eero.Hyvonen@cs.helsinki.fi; eero.hyvonen@tkk.fi

Mike Jackson Centre for Geospatial Science, University of Nottingham, Nottingham, UK, mike.jackson@nottingham.ac.uk

Prateek Jain Kno.e.sis Center, Wright State University, Dayton, OH 45435, USA, jainprateek@gmail.com; prateek@knoesis.org

Tomi Kauppinen Institute for Geoinformatics, University of Münster, Münster, Germany, tomi.kauppinen@uni-muenster.de

Carsten Keßler Institute for Geoinformatics, University of Münster, Münster, Germany, carsten.kessler@uni-muenster.de

Werner Kuhn Institute for Geoinformatics, University of Münster, Münster, Germany, werner.kuhn@gmail.com; kuhn@uni-muenster.de

Jens Ortmann Institute for Geoinformatics, University of Münster, Münster, Germany, j_ortm02@uni-muenster.de

Matthew Perry Oracle, 1 Oracle Drive, Nashua, NH 03062, USA, matthew.perry@oracle.com

Robert G. Raskin Science Data Systems Section, NASA/Jet Propulsion Laboratory, Pasadena, CA, USA, robert.g.raskin@jpl.nasa.gov

Simon Scheider Institute for Geoinformatics, University of Münster, Münster, Germany, simon.scheider@uni-muenster.de

Amit P. Sheth Kno.e.sis Center, Wright State University, Dayton, OH 45435, USA, amit.sheth@wright.edu; amit@knoesis.org

Thomas Shoberg United States Geological Survey, Rolla, MO, USA, tshoberg@usgs.gov

Kristin Stock Centre for Geospatial Science, University of Nottingham, Nottingham, UK, Kristin.Stock@nottingham.ac.uk

Johannes Trame Institute for Geoinformatics, University of Münster, Münster, Germany, johannestrame@uni-muenster.de

Jouni Tuominen University of Helsinki, Helsinki, Finland, jouni.tuominen@cs.helsinki.fi

E. Lynn Usery United States Geological Survey, Rolla, MO, USA, usery@usgs.gov

Jari Vaatainen Geological Survey of Finland, Espoo, Finland, jari.vaatainen@gtk.fi

Dalia Varanka United States Geological Survey, Rolla, MO, USA, dvaranka@usgs.gov

Chapter 1
Representing and Utilizing Changing Historical Places as an Ontology Time Series

Eero Hyvönen, Jouni Tuominen, Tomi Kauppinen, and Jari Väätäinen

Abstract Place names and their geographical coverage change in time. This causes problems when retrieving information content related to different times. Geo-content is usually indexed using place names of the time of indexing (e.g. a photo of the 1968 upraise of Czechoslovakia indexed then) or of the time that the content has been used or created (e.g. a spear used in the Punic Wars in 146 B.C. in Carthago but indexed at a later time using place names of that time). Finally, end-users may query content in terms of contemporary place names (e.g. Check Republic or Slovakia) or overlapping historic names of different times (e.g. Roman Empire). This chapter presents an ontology-based approach to this problem. The idea is to represent and maintain a time series of spatial ontologies in terms of easily manageable local spatio-temporal changes from which the actual time series ontology can be generated automatically with semantic enrichment. This ontology can then be used for indexing and for mapping spatio-temporal regions and their names onto each other. As a proof-of-concept, the system has been applied to modeling the history municipalities of Finland in 1865–2010. We present the model, a tool for maintaining the change history in a user-friendly way, transformation of the place change history into an ontology time series with semantic enrichment, and publication of the ontology as a ready to use ontology services on the web with AJAX, Web Service, and REST interfaces. The system has been applied in the semantic cultural heritage portal CULTURESAMPO for semantic search and recommendation, as well as an external service for indexing cultural heritage content, and for query expansion search in a legacy cultural heritage database system.

E. Hyvönen (✉)
Aalto University, Aalto, Finland
e-mail: eero.hyvonen@tkk.fi; Eero.Hyvonen@cs.helsinki.fi

N. Ashish and A.P. Sheth (eds.), *Geospatial Semantics and the Semantic Web:
Foundations, Algorithms, and Applications*, Semantic Web and Beyond 12,
DOI 10.1007/978-1-4419-9446-2_1, © Springer Science+Business Media, LLC 2011

1.1   Introduction

Metadata on the Semantic Web is based on referencing to concepts of ontologies [26, 34]. There are lots of databases and repositories available for current places, such as GeoNames.[1] Dealing with historical geographical content adds the temporal dimension and notion of change to geographic information systems (GIS). For example, a reference to "Germany" or "the U.S." may refer to different regions (e.g. Germany in 1943 vs. 1968), depending on the time of reference.

There are vocabularies and ontologies describing historical places, such as the Thesaurus of Geographical Names (TGN).[2] From a geographical viewpoint, such vocabularies typically tell the part-of hierarchy of places, and a coordinate point of the place or its polygonal area, various metadata for human users, and an identifier for referencing the concept. For example, in TGN the entry for the city 'New York' list its various names, such as 'New Amsterdam' and 'Big Apple', tells its hierarchic position in the U.S. (e.g. that it belongs to the state of New York) and additional larger regions, place types (e.g. city, port, national capital in 1778, etc.) and references to literal and other sources explaining e.g. the alternative names, such as 'New Amsterdam' (historical place) in more detail.

1.1.1   Limitations of Historical Geo-vocabularies

If content is annotated with a current or a historical place name and queried with the same name, stored content can be found. However, names have multiple meanings (e.g. Paris in France vs. Paris in Texas) and places can be annotated and referred to using geographically overlapping concepts with different names. In a time perspective, a region R can be referred to in principle by any region name at different granularity levels that has at some point of time overlapped R. For example, Helsinki in Finland, can be referred to by any regional boundaries of the city since its establishment in 1550, by the various incarnations of the neighboring regions annexed to Helsinki, by different regions of Sweden before the Napoleonic wars, by Russian regions in the nineteenth century, by regions of independent Finland since 1917, and by EU nomenclature since 1995. A simple approach used e.g. in TGN is to associate names with alternative names, but this is problematic when the same area or its part can be referred to by *different* overlapping places. A part-of hierarchy eases the pain w.r.t. regions and subregions, but even then there is the problem that the hierarchy is time dependent. For example, New Amsterdam has been part of the Netherlands, but is used as an alternative name for contemporary New York in TGN. The city was renamed 'New York' only in 1664 by the Duke

---

[1] http://geonames.org/.

[2] http://www.getty.edu/research/conducting_research/vocabularies/tgn/.

of York under the British rule. Also many other relations of regions change in time. For example, New York used to be the capital of the U.S. but is not any more.

For a more accurate and machine interpretable representation of historical places, the notion of a spatio-temporal named region during a period of time is need. Relating such regions or places ontologically with each other is needed in information retrieval, because the end-user may not use the same place names in search queries that are used in annotations, but only related place names. More generally, ontological, topological and other relations between historical places are needed in order to link semantically related content with each other in applications, such as recommending systems and semantic portals of cultural heritage [16].

1.1.2   Research Questions

From the perspective of the Semantic Web, this need creates new research questions, such as:

- **Spatio-temporal ontology models** How to represent geo-ontologies of spatio-temporal places that change in time?
- **Spatio-temporal ontology maintenance** How to maintain spatio-temporal ontologies that change in time?
- **Annotation support** How to support content creation using such ontologies, so that correct references to places in time can be made?
- **Application** How to utilize such spatio-temporal ontologies in applications for e.g. querying, recommending, content aggregation, and visualization?

1.1.3   Chapter Outline

In this chapter an approach is presented addressing these research questions. We first formulate a model for representing spatio-temporal regions as an ontology time series. We present methods for creating such ontologies based on geographical changes and incomplete data – a typical situation when dealing with historical places. This part of the chapter is based on and presents an overview of a series of papers published by the authors earlier, especially [23, 25], with some extensions. In particular, we emphasize aspects related to creating historical geo-ontologies based on incomplete knowledge. After this an ontology service is presented by which the ontology can be published easily and used in external legacy systems and applications as a service [39]. Two applications of the ontology are discussed: a semantic portal for cultural heritage [18] and a query expansion service [40] attached to a legacy application on the web.

The work is part of the national FinnONTO project (2003–2012)[3] aiming a building a national semantic web infrastructure [19].

1.2   A Model for Spatio-Temporal Ontology Time Series

Major goals and motivations for developing the spatio-temporal ontology model are:

1. **Accurate annotations.** Facilitate more accurate content descriptions in metadata using spatio-temporal regions.
2. **Semantic search.** Facilitate search by query or document expansion in applications, based on spatio-temporal relations.
3. **Semantic linking.** Facilitate finding and aggregating related content in applications, based on spatio-temporal relations.
4. **Semantic enrichment.** Facilitate enriching of the ontology automatically by reasoning. A human developer does not need to describe everything explicitly in the ontology, but part of the properties and relations can be created by the machine based on the semantics.
5. **Visualization.** Facilitate using ontological structures in user interfaces, e.g. the part of hierarchy at different times.

To achieve these goals, spatio-temporal regions and their collections are used as annotation concepts with persistent URIs, and are defined and related to each other by a time series of ontologies. We focus on representing *spatio-temporal regions* (STR). "Region" is a commonly used geographic term in different branches of geography. Regions can be defined based on various features and include e.g., political, religious, natural resource, and historical regions.[4]

Regions of different kinds can be characterized from a spatio-temporal point of view by the following core properties: name, time span, size, and polygonal area. Regions can be related with other by topological relations [5], such as

1. The part-of relation defining hierarchies.
2. Overlap relation telling how much regions overlap.
3. Other relations, such as neighbor-of, near-by etc.

These relations are potentially useful in query expansion [3, 20] and in semantic linking on a spatial dimension. For example, when searching for castles in Europe, it makes sense to return castles in different countries that are part of Europe. However, from an IR query expansion point of view, it is not always clear when the relations can be used. For example, when querying documents about the EU, one probably is not so interested in documents about the member states but documents about the EU as a whole. Here recall is enhanced but at the cost of precision.

---

[3]http://www.seco.tkk.fi/projects/finnonto/.
[4]http://en.wikipedia.org/wiki/Region.

In this chapter we assume that the ontology is applied wisely in situations where utilizing a relation matches the needs of the application case.

In below, our spatio-temporal ontology model is first outlined, and after this the problem of creating it from partial geographical data available.

1.2.1   A Model of Ontology Time Series

A major reasoning task in our ontology model is to compute the overlap relation between the regions in an ontology. This relation is represented by the properties *overlaps* (covers) and its reverse *overlappedBy* (coveredBy). Assume that the area of a region A is 100, the area of B is 200, and that the shared common area C of A and B is 50. Then A overlaps B by $C/B = 0.25$ and B overlaps A by $C/A = 0.5$. If a query uses the concept A that overlaps B, then content annotated using B could be returned and the hits can be sorted in the order of relevance based on the degree of overlap (here 0.25). On a temporal dimension, regions can be related through the overlap of their co-existence in time.

In an ideal situation, the polygons of the regions in an ontology are known. Then the overlap relation between all pairs of regions can be computed straightforwardly. Furthermore, based on polygons of regions, additional topological relations, such as neighbor-of, east-of etc. can be reasoned/computed, and the ontology be enriched. However, a key problem here in practice is that polygon data is not always available, which is especially common when dealing with historical places. In many cases the polygon of a region may not even be known or its boundaries are uncertain. Then one has to start ontology creation from what data is available, enrich the knowledge by whatever means are available, and be content with a final partial model, too. A major benefit of using ontologies for representing spatio-temporal regions is that semantics enable automatic enrichment of human input knowledge, saving time and money in content creation, and facilitating implementation of more "intelligent" applications.

The central concept in our ontology model is the STR. It has three core properties: (1) a name by which the region is referred to, (2) a bounded geographical polygonal area, and (3) a time interval that the region with the name existed without change w.r.t. name and time. Each spatio-temporal region has an identity of its own and is labeled as: *placename(begin, end)*. For example, 'Helsinki (1931–1945)' refers to the region of Helsinki from 1931 to 1945. Depending on the application, an STR has additional spatio-temporal properties and semantic relations with other spatio-temporal regions, such as size, part-of, neighbor-of etc., and domain specific properties, such as population, main religion, natural resource type, etc.

A collection of spatio-temporal regions with the same place name can constitute a *spatio-temporal spaceworm* that essentially defines a region over time. For example, the city of 'Helsinki' as an administrative area can be defined as a spaceworm defined by its constituents: 'Helsinki (1550–1639)', 'Helsinki (1640–1642)', 'Helsinki (1643–1905)', 'Helsinki (1906–1911)', 'Helsinki (1912–1926)',

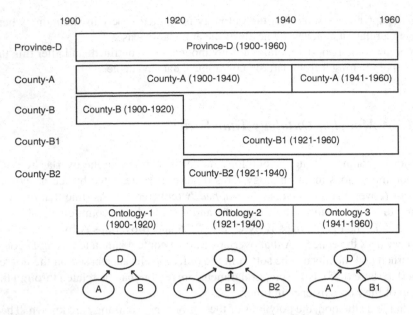

Fig. 1.1 Example of an ontology time series based on two regional changes

'Helsinki (1927–1930)', 'Helsinki (1931–1945)', 'Helsinki (1946–1965)', 'Helsinki (1966–2008)', and 'Helsinki (2009–)'. The region of Helsinki is defined by the union of these STRs.

The ontology in our model is essentially defined as a set of STRs and space-worms. At each moment t the world consists of the regions $\{placename(x,y)|x \leq t \leq y\}$. Therefore, at any point in time t when a region change takes place, i.e. when at least one STR is created ($placename(t,x)$) or vanishes ($placename(x,t)$), a different new set of STRs defines a *period ontology O* describing the world until the next change.

A period ontology is characterized by the properties of its regions. The relations between the regions that can be defined according to the application needs. In our case ontology for the Finnish historical municipalities (to be presented later), for example, we represent countries, provinces, and municipalities as STRs. A country is divided exhaustively into a set of provinces, and each province into a set municipalities using the hierarchic part-of relation.

The temporal sequence of period ontologies defines an ontology time series. It is intuitively a sequence of partonomies. Each period ontology is valid between two nearest subsequent changes. However, STRs in the partonomies are related with each other globally by the overlap relation. If two regions do not overlap, the degree of overlap is 0, a value in (0,1) is used if they share area, and value 1 means a total coverage.

For example, Fig. 1.1 depicts a situation, where a province D that consists of two counties A and B is established at 1900. County B is split into two counties

B1 and B2 on January 1 in 1921, and on January 1 in 1941 county B2 is merged into A. The spaceworms of the provinces and the counties involved are depicted as horizontal boxes in a row stretching over time. For example, spaceworm A has two constituents. The graph tells the following story: County B vanishes as a result of a split into counties B1 and B2 in 1921. In 1941, B2 vanishes, because it is merged into A. At this point a new constituent is created for A because of the change in the area of the region A, but the new incarnation 'County-A (1941–1960)' is still a member of the spaceworm of County-A because B2 merges into A without changing the name of A. In the lower part of the figure, the part-of hierarchy of each period ontology is visualized as an ontology time series. Here shorthand node labels A, B, B1, and B2 refer to the corresponding STRs above, and A' to 'County-A (1941–1960)' that includes the region of B2.

The ontology time series is used for annotating content by spatio-temporal regions, when dealing with temporal materials. For example, a film about Helsinki during the Winter War in 1939 would be annotated by the resource 'Helsinki (1931–1945)'. When a generic reference to a region is made without considering the time dimension, the spaceworm resource can be used, e.g. when annotating a book about Helsinki at different times. The major benefit of using the ontology is that resources in annotations are now more accurate (e.g., modern Helsinki covers a much wider area than the historical versions of Helsinki), they can be associated with time, and they can be related with each other through the part-of, overlap and other relations. This facilitates query expansion and semantic linking of regions even if their names are different.

1.2.2   Enriching the Ontology

A major benefit of the model outlined above is that the ontology can be enriched semantically using reasoning. This can be especially useful when only partial or inexact knowledge about places is available, which is typical when dealing with historical data. Uncertainty may be related to any core property of an STR: name, area, and time. In the following, we focus on the problem of dealing with incomplete information about the polygonal areas and spatial relations of STRs. For representing uncertainty in names, properties such as skos:altLabel or skos:hiddenLabel of the SKOS vocabulary standard[5] can be used. A way to represent uncertainty in interval end-points is to use four-point intervals, as suggested e.g. in the CIDOC-CRM standard.[6]

If historical documents do not specify the geographical boundaries of a region, qualitative information about spatial changes may still be available. In our case study [25], for example, polygons of older incarnations of municipalities were not

---

[5]http://www.w3.org/2004/02/skos/.
[6]http://www.cidoc-crm.org/.

Fig. 1.2 Overlap relation based on the changes of the ontology of Fig. 1.1, and known areas of the regions listed in the leftmost column. A' refers to County-A after the merge

	D	A	A'	B	B1	B2
D=100	1	1	1	1	1	1
A=40	40/100	1	40/60	0	0	0
A'=60	60/100	40/60	1	20/60	0	1
B=60	60/100	0	20/60	1	1	1
B1=40	40/100	0	0	40/60	1	0
B2=20	20/100	0	20/60	20/60	0	1

available (or digitization was not possible), but usually the sizes of the areas (in km^2) and change events, such as emergence of a new county by merging two old ones at a certain year, were known. We therefore postulated that a spatio-temporal ontology, as described above, has to be created based on several datasets that may be more or less complete when starting ontology creation:

1. Repository of regions (R) defining the name, type, size, and time interval of STRs, and application specific features.
2. Repository of regional changes (RC): explicit information about how regions e.g. are established, vanish, split, and merged.
3. Repository of polygons of regions (PR): the coverage of STRs.
4. Repository of topological relations between STRs (TR): additional relations between STRs, as needed in applications.

The final RDF ontology consists of an union of these components enriched by additional triples generated by reasoning. Let us assume that R is fully specified. Then the ontology can be enriched as follows:

1. Time series. Based on R, the ontology time series can be generated by splitting the time line at each STR interval limit, and collecting overlapping STRs into period ontologies.
2. Based on RC and PR, additional polygons in PR can be generated. For example, the polygon of a merged STR is the union of the polygons of its constituents.
3. Based on RC and PR, topological relations can be generated.

As an example of generating topological relations in this framework, Kauppinen and Hyvönen [23] presents a method for determining the overlap relation between STRs based on R and RC. The result is basically a *regions* × *regions* matrix defining the degree of overlap relation between all pairs of regions: given a region its overlaps w.r.t. other region can be read from the corresponding row in the table instantly. The relation was can be populated into the RDF base as a set of *overlaps* property triples, or its inverse *overlappedBy*.

For example, given the RC illustrated in Fig. 1.1, the overlap table of Fig. 1.2 can be computed. On the leftmost column the areas of the STRs in Fig. 1.1 are given. For example, since B (area 60) is split into B1 (40) and B2 (20), B2 overlaps B by their shared area, i.e. by $20/60 = 1/3$.

1.3   Case Study: Historical Finnish Municipalities

The model and methods described in the previous sections were applied to create the Finnish Spatio-temporal Ontology SAPO,[7] an ontology time series of Finnish municipalities over the time interval 1865–2007 [25]. Also since 2007, the model has been kept in concordance with later changes of administrational regions and municipalities in Finland. Most Finnish municipalities have overcome some kind of areal changes, many of them several times after their establishment. Figure 1.3 shows in dark color municipalities that haven't had any changes since 1865 [24].

SAPO is an instance of the general problem of modeling boundary changes of provinces, municipalities, and other regions in different countries. For example in Japan the number of municipalities has declined from about 71,000 in 1889, to about 1,700 in 2008 [2]. During this period many old municipal names were dissolved, and various new names were generated. In Japan, from the year 1999 until 2008, a total of 598 municipalities were formed by merging existing ones, out of which 330 kept their existing names and 268 got new names.

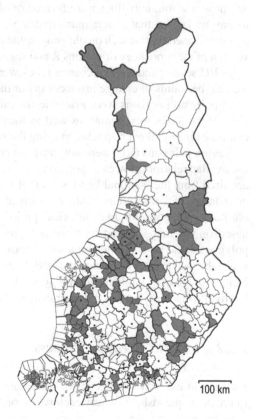

Fig. 1.3 Regional changes are common in Finland: *dark color* indicates municipalities whose name or area has not changed since 1865. Courtesy of the National Land Survey of Finland

100 km

[7]http://www.seco.tkk.fi/ontologies/sapo/.

Table 1.1 Different types of regional changes of municipalities between 1865 and 2007 in Finland

Change type	Quantity
Establishment (A region is established)	508
Merge (Several regions are merged into one)	144
Split (A region is split to several regions)	94
Namechange (A region changes its name)	33
Changepartof (Annexed (to a different country))	66
Changepartof (Annexed (from a different country))	1
Changepartof (Region moved to another city or municipality)	256
Total sum	1,102

1.3.1 Developing the Ontology

In our case, the information available in the outset was lists of municipalities at different times telling e.g. the areas of the regions, to which province they belonged, and how new municipalities were formed or old ones were changed. For example, it may be known that a new municipality was formed by merging two old ones together. Based on research on old geographical books, lists, and other data, the first version of the repository of regions R and regional changes RC could be created.

In RC seven fundamental change types were identified. Table 1.1 lists them as well as the counts of change instances in our dataset (in 2007):

Initially no polygons were available for calculating the overlaps. However, the sizes of the STRs were known as well as local changes, which made it possible to compute the global overlap relation using the model and methods discussed above.

Region polygons (RP) were not available and therefore not used in determining the overlap relation. However, polygons for contemporary municipalities were later acquired from the National Land Survey of Finland, and in old maps geographical boundaries of some areas could be seen at certain time points. To enrich the ontology, polygons for two historical period ontologies were digitized by hand based on old maps. Based on these polygons and the change history, additional polygons could be computed by a set of reasoning rules. After this, the time series was published as a service using the ONKI ontology service [41]. A large amount of content in the final published ontology has not been created by a human ontologist but by the machine, based on the semantics of the ontology.

1.3.2 Content Creation Process

An easy to use way to encode the information about regional changes (RC) was to create a spreadsheet, where each row represents a spatio-temporal change. The columns represent the properties of the changes, such as the type of the change, time, and regions involved, implementing the metadata schema for regional changes.

Place	Time	Change	From 1	From 2	Moved parts	Where 1	Where 2	Description
Viiala	2007	Merged	Viiala	Toijala		Akaa_2		Toijala ja Viiala yh
Viipuri	1403	Addition				Viipuri		Viipuri sai kaupunk
Viipuri	19.09.1944	Changepartof	Viipuri			Russia		Luovutetuilla aluei
Viipurin mlk	1869	Addition				Viipurin mlk		
Viipurin mlk	1906	Split	Viipurin mlk			Nuijamaa	Viipurin mlk	Nuijamaa itsenäist
Viipurin mlk	1921	Split	Viipurin mlk			Vahviala	Viipurin mlk	
Viipurin mlk	19.9.1944	Changepartof	Viipurin mlk			Russia		Luovutetuilla aluei
Viitasaari	1868	Addition				Viitasaari		Viitasaari perustetti
Viitasaari	1934	Changepartof	Viitasaari		Vuoksenkoski	Kannonkoski		Viitasaaresta erotet
Viljakkala	1874	Addition				Viljakkala		Perustettu 1874
Viljakkala	2007	Merged	Viljakkala	Ylöjärvi		Ylöjärvi		Viljakkala yhdistyi
Vilppula	1904	Changepartof	Keuruu		Suuri osa Keuruunk	Vilppula		
Vilppula	1904	Changepartof	Ruovesi		"osia"	Vilppula		
Vilppula	1922	Split	Vilppula			Mänttä	Vilppula	Mänttä erosi Vilppu
Vilppula	1954	Changepartof	Kuorevesi		Kuoreniemi, osa Aj»	Vilppula		Kuorevedestä liitet
Vilppula	1973	Split	Pohjaslahti			Vilppula	Virrat	Jaettu 1973 (Vilppu
Vimpeli	1866	Addition				Vimpeli		Perustettu 1866
Virolahti	1908	Changepartof	Virolahti		Heikkilä, Järvelä, Jo	Miehikkälä		1908 Virolahdesta
Virolahti	19.9.1944	Changepartof	Virolahti		"osia"	Russia		
Virrat	1868	Addition				Virrat		Perustettu 1868
Virrat	1973	Split	Pohjaslahti			Vilppula	Virrat	Jaettu 1973 (Vilppu
Virtasalmi	1912	Split	Pieksämäki			Pieksämäki	Virtasalmi	
Virtasalmi	2004	Merged	Jäppilä	Pieksämä»		Pieksänmaa		
Vuoksela	1914	ChangepartofAddition			Muolaa: Vuosalmi	Vuoksela		Perustettu 1914 (V.

Fig. 1.4 Maintaining SAPO-ontology as a spreadsheet table

Figure 1.4 shows a screenshot of the metadata of changes. Different schema fields, such as 'Place', 'Date', 'Change' (type), and 'Moved parts', are represented as columns, and are filled up with unique references to resources or with other values. STRs are referred to by their names (including the time interval). For example, the split of 'Viipurin mlk (1869–1905)' into 'Nuijamaa (1906–1944)' and 'Viipurin mlk (1906–1920)' is seen on the row 1194, and the annexing of 'Viipurin mlk' from Finland to Russia on 1944–09–19 is on the row 1196. Most changes have also a natural language explanation of the event for human users.

The process from the spreadsheet, maintained by a human cataloger, to the publication of the ontology time series proceeds in the following steps:

1. The spreadsheet is saved in CSV format.
2. A script transforms the CSV form into RDF.
3. Overlap relations of spatio-temporal regions are computed as explained above, and represented as properties of the regions.
4. Additional information concerning the metadata can be added to the knowledge base, such as boundaries of regions as polygons at certain points of time.
5. The ontology is enriched further by reasoning new polygons based on known polygons and the change history.
6. The ontology is enriched further by reasoning additional topological relations between the STRs, e.g. that two municipalities are neighbors.
7. The ontology time series is generated from the change history, one period ontology for each two subsequent changes.
8. The time series is published using ONKI ontology service (to be explained in more detail below).

The methods for enriching and creating an ontology time series from the spreadsheet CSV metadata were implemented using Java and Jena Semantic Web

Framework.[8] The resulting RDF repository contains 1105 different changes and 976 different STRs of 616 different historical and modern places (spaceworm), meaning that each place has on average 1.58 temporal parts. For example, the spaceworm resource 'Viipurin mlk' includes the STRs 'Viipurin mlk (1869–1905)', 'Viipurin mlk (1906–1920)', 'Viipurin mlk (1921–1943)', and 'Viipurin mlk (1944–)'. The temporal parts and their partonomy hierarchies in the RDF repository constitute 142 different temporal period ontologies between the years 1865 and 2007, each of which is a valid model of the country during its own time span.

1.4   Publishing the Ontology as an ONKI Service

The ONKI Ontology Service [41] is a general ontology library that acts as a publishing channel for ontologies and provides functionalities for accessing them using ready-to-use web widgets as well as APIs for both humans and machines. ONKI supports services such as content indexing, concept disambiguation, searching, and (URI) fetching. The service is based on ontology and domain specific implementations of ONKI servers which conform to the ONKI application interface [42]. This means that it is possible to provide a single web widget to access all ontologies, and at the same time, provide domain-specific user interfaces and technical implementations optimized for ontologies of different sizes, modeling languages and principles.

ONKI SKOS [39] is an ontology server supporting thesaurus-like ontologies especially in content indexing. ONKI SKOS can be used to browse, search and visualize any vocabulary conforming to the SKOS recommendation, and also RDF(S) and OWL ontologies with additional configuration. ONKI SKOS does simple reasoning, e.g. transitive closure over class and part-of hierarchies. The implementation has been tested using various ontologies, such as the Finnish Spatio-temporal Ontology SAPO.

ONKI SKOS Browser (see Fig. 1.5) is the graphical user interface of the ONKI SKOS server. It consists of three main components: (1) *concept search with semantic autocompletion*, (2) *concept hierarchy*, and (3) *concept properties*. When typing text to the search field, a query is performed to match the concepts' labels. The result list shows the matching concepts, which can be selected for further examination. The search can be further narrowed by restricting the search to concepts of a certain type or to a desired subtree of the ontology. When a concept is selected, its concept hierarchy is visualized as a tree structure, and its properties are shown as a table.

In Fig. 1.5 user has searched all the temporal municipalities whose name starts with a string "helsinki", referring the spaceworm 'Helsinki'. Matching STRs are shown, after each input character, as a list of choices on the left. In this case, the

---

[8]http://jena.sourceforge.net/.

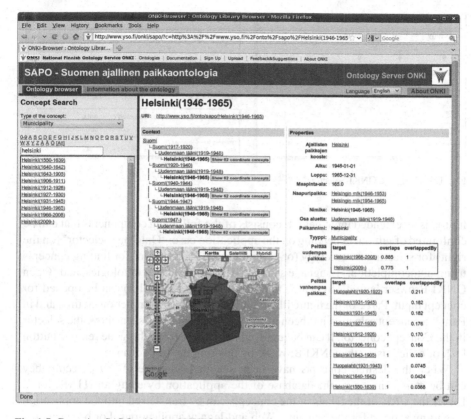

Fig. 1.5 Browsing SAPO with the ONKI SKOS Browser

user has already selected the STR 'Helsinki (1946–1965)' for inspection and visualization. The part-of relations of the STR are shown as a hierarchy tree on the right – 'Helsinki (1946–1965)' is part of the province 'Uudenmaan lääni (1919–1948)', which is part of several spatio-temporal incarnations of the country Finland. The geographical region of the place is shown as a polygon on a Google Maps[9] view. On the right hand side, neighbouring and overlapping municipalities are shown. For example, 'Helsinki (1946–1965)' overlaps 'Huopalahti (1920–1922)' with a weight 1, since Huopalahti has been annexed to Helsinki.

The ONKI Ontology Services can be integrated as mash-ups into applications on the user interface level (in HTML) by utilizing the ONKI Selector, a lightweight web widget providing functionalities for accessing ontologies, e.g., for content annotation purposes. The ONKI Selector depicted in Fig. 1.6 can be used to search and browse ontologies, fetch URI references and labels of desired concepts, and to store them in a concept collector in HTML code. The selector, depicted in Part 1 of the

---

[9]http://maps.google.com/.

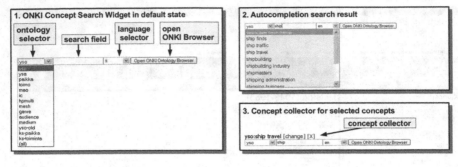

Fig. 1.6 Using the ONKI Selector

figure, is an extended input field. It consists of the following components that can be configured of left out depending on the application case: 'Ontology selector' (on the right) for selecting an ontology (or several ones),'Search field' for finding concepts using autocompletion, 'Language selector' for multi-lingual ontologies, and 'Open ONKI Browser button', by which the ONKI Browser (Fig. 1.5) can be opened for concept input. Part 2 of the figure illustrates using the autocompletion facility, and in Part 3, a concept selection has been made, and the concept is seen above the selector in the Concept collector. It can be removed from there by pushing the remove button [×], or edited using the ONKI Browser by pushing the link 'change'.

When the desired concepts have been selected with the ONKI Selector they can be stored into, e.g., the database of the application by using an HTML form. Either the URIs or the labels of the concepts can be transferred into the application providing support for the Semantic Web and legacy applications. For browsing the context of the concepts in ontologies, the ONKI SKOS Browser can be opened by pressing a button. Once suitable concepts are found, they can be fetched from the browser to the application.

ONKI Ontology Service provides for machine usage APIs which can be used for, e.g., querying for concepts by label matching, getting properties of a concepts, and getting metadata about an ontology. The ONKI API has been implemented in three ways: as an AJAX service, as a Web Service, and a simple HTTP API.

1.5 Applications

1.5.1 CultureSampo

SAPO ontology is in use in the semantic portal "CULTURESAMPO– Finnish Culture on the Semantic Web 2.0"[10] [18] that contains hundreds of thousands of cultural

---

[10]http://www.kulttuurisampo.fi/.

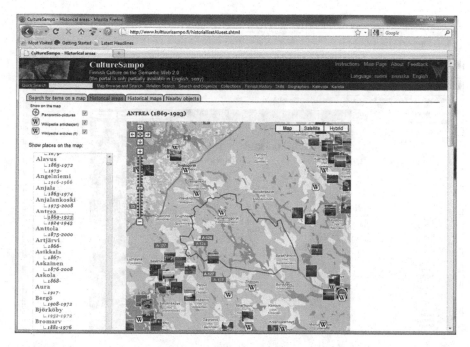

Fig. 1.7 The user has selected historical 'Antrea (1869–1923)' on the *left*, and the area is shown on the map with articles from Wikipedia and photos from Panoramio

heritage content items of different kinds from different organizations and the public. The systems uses SAPO ontology for providing the end-user with following functionalities:

1. Old places are of interest of their own – just knowing where and when they existed is already valuable. In CULTURESAMPO, old places of SAPO can be found as an index; by clicking on a name, the area is shown on a map with other content. For example, in Fig. 1.7 the user has selected the historical municipality of 'Antrea (1869–1923)' in the index on the left, and the system shows its boundaries on the map.
2. Information based on coordinates can be associated with regions by showing them simply on a map, as customary in traditional Google Maps applications. In Fig. 1.7, links to contemporary datasets are provided on maps, in this case Wikipedia articles and Panoramio[11] photos related to the area. In CULTURE-SAMPO also modern places that are inside the polygonal boundaries of the historical region can be retrieved, and can be used to browse the map (this feature is not seen in the figure). For modern places the ONKI-Geo [17] ontology service is used.

---

[11]http://www.panoramio.com/.

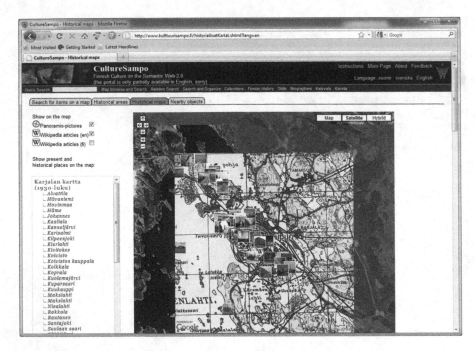

Fig. 1.8 Old maps overlayed transparently over contemporary maps and satellite images show historical changes

3. STRs can be used as a basis for semantic recommending, based on the metadata such as time and topological relations. In Fig. 1.8, the user has selected to view the STR 'Viipuri (1920–1944)'. The system shows content related to it through semantic associations, including folk poems, music, artifacts, paintings etc. Figure 1.9 shown these recommendations as symbol links; these recommendations can be found in the view of Fig. 1.8 under the map (scrolling down is needed). Also content from historical regions that overlap 'Viipuri (1920–1944)' are listed as recommendations. The overlaps are based the global overlap table derived from the change history of municipalities. In recommending, the CULTURESAMPO knowledge base is used as a SPARQL end-point.

4. Visualization of historical changes. Figure 1.8 depicts the Temp-O-Map system [22] in CULTURESAMPO that utilizes the ontology time series in visualizing historical and modern regions on top of maps and satellite images. Historical places, i.e. STRs, can be selected from a drop-down menu on the left. Here the temporal constituent 'Viipuri (1920–1944)' of 'Viipuri' is selected. By viewing old and contemporary maps on top of each other gives the user better understanding about the history of the region. In this case the Viipuri area was annexed to the Soviet Union after the World War II, and many old Finnish place names were changed in new Russian ones and are also now written using Cyrillic alphabet. In the middle, a smaller rectangular area is shown with a

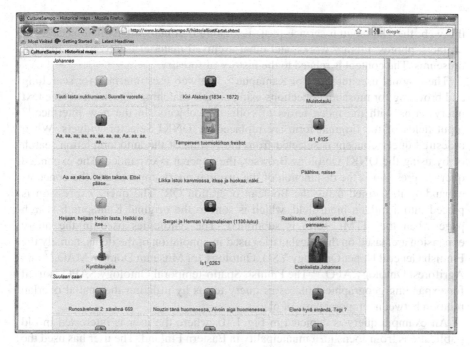

Fig. 1.9 Semantic recommendation links related to 'Viipuri (1920–1944)'

semi-transparent[12] old Karelian map that is positioned correctly and is of the same scale as the Google Maps image. In order to move around the user is able to use the zooming and navigation functions of Google Maps and the historical view is automatically scaled and positioned accordingly.

To provide the historical maps, we used a set of old Finnish maps from the early twentieth century covering the area of the annexed Karelia region before the World War II. The maps were digitized and provided by the National Land Survey of Finland.[13] In addition, a geological map of the Espoo City region in 1909, provided by the Geological Survey of Finland,[14] was used. This application is also included in the CULTURESAMPO portal.

1.5.2 Semantic Query Expansion Service

For demonstrating the utilization of ontology services in query expansion, we extended the ONKI Selector widget with functions for expanding input queries,

---

[12]We use transparency libraries provided by http://www.kokogiak.com/ which allow the alteration of the level of transparency.

[13]http://www.maanmittauslaitos.fi/default.asp?site=3.

[14]http://en.gtk.fi.

and integrated it with the search interface of an existing legacy search system on the web, the Kantapuu.fi service [40]. Kantapuu.fi contains tens of thousands of artifacts, photos, literary works, and other archived material from various Finnish museums. The content is related to the history of forestry.

The original user interface of Kantapuu[15] is a web user interface for searching and browsing for museum collections using simple matching algorithm of free text query terms with the index terms of collection objects. In the new interface,[16] input fields of the original form are replaced by ONKI Selector widgets. When a desired query concept is selected from the results of the autocompletion search or by using the ONKI Ontology Browser, the concept is expanded. The expanded query expression is the disjunction of the original query concept and the concepts expanding it, formed using the Boolean operation OR. The query expression is placed into a hidden input field, which is sent to the original Kantapuu.fi search page when the HTML form is submitted. The ontologies used in the query expansion are based on the vocabularies used in annotation of the items, namely the Finnish General Upper Ontology YSO, Ontology for Museum Domain MAO,[17] and Agriforest Ontology AFO.[18] The Finnish Spatio-temporal Ontology SAPO is used for expanding geographical places as query terms by utilizing the spatial overlap relation between temporal parts of places.

An example query is depicted in Fig. 1.10, where the user is interested in old publications from Joensuu, a municipality in Eastern Finland. The user has used the autocompletion feature of the widget to input to the *keywords* field the query term "publicat". This string has been autocompleted to the concept *publications*, which has been further expanded to its subclasses (their Finnish labels), such as *books*. Similarly, the place spaceworm *Joensuu* has been added to the field *place of usage* and expanded with the STRs it overlaps.

The result set of the search contains four items, from which two are magazines used in Eno (a municipality overlapping Joensuu) and the rest two are cabinets for books used in Joensuu. Without using the query expansion the result set would have been empty, as the place *Eno* and the concept *books* were not in the original query.

Expanding queries using the spatial overlap relation between places is often useful for enhancing recall, but may decrease the precision of the query by introducing irrelevant query terms. For example, if a user is interested in historical items found in a place A, which overlaps a place B only a little, he may not appreciate search results concerning items found in the parts of the place B that do not overlap the place A and are far from it. To manage situations like these, query expansion has been made transparent to the user. The user

---

[15]http://www.kantapuu.fi/, follow the navigation link "Kuvahaku".

[16]A demonstration is available at http://www.yso.fi/kantapuu-qe/.

[17]http://www.seco.tkk.fi/ontologies/mao/.

[18]http://www.seco.tkk.fi/ontologies/afo/.

1. Search query

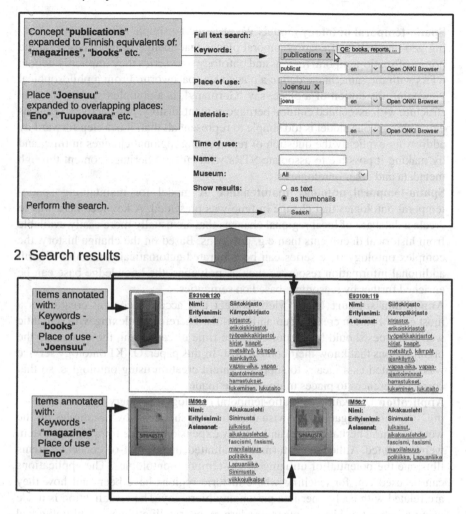

Fig. 1.10 Kantapuu.fi system with integrated ONKI widgets

is always able to view the expansion, select whether to use query expansion or not, and remove the suggested query expansion concepts from the query if needed.

1.6 Discussion

In conclusion, we briefly review answers to the research questions set in Sect. 1.1.2, discuss related work, and outline directions for further research.

1.6.1   Research Questions Revisited

- **Spatio-temporal ontology models** We presented a simple model for representing geo-ontologies of spatio-temporal places that change in time, based on the notion of spatio-temporal regions and ontology time series they implicitly define. STRs with the same name define a place as a spaceworm. From a philosophical viewpoint, the notion of a place, say 'Germany', is a complex spatio-temporal structure with associated cultural heritage content, history, perspectives, opinions etc. Although our model is too simple to represent all that, it is a step forward by addressing explicitly the question of representing regional changes in time, and by making it possible to associate STRs with cultural heritage content through metadata and other ontologies.
- **Spatio-temporal ontology maintenance** A model for maintaining spatio-temporal ontologies that change in time was presented. A key idea here was to create a database of local regional changes that are usually more easily available from historical documents than e.g. polygons. Based on the change history, the complex ontology time series can be generated automatically. Combined with additional information resources such as polygons, the knowledge base can be enriched further by reasoning, based on semantics.
- **Annotation support** In our view, correct and accurate content creation is a most critical part in creating semantic portals. Therefore, indexing with semantic web resources should be supported at the time of cataloging the content in the organizations that know their content best. In this paper, ONKI ontology service was presented as a means to support content creation using ontologies, so that correct references to places in time can be made.
- **Application** Utilization of spatio-temporal ontologies in querying, recommending, content aggregation, and visualization was shown by two examples on the web: a cultural heritage portal and a query expansion service for a legacy system were presented. Although not formally evaluated, these proof-of-concept systems illustrate the potential of utilizing spatio-temporal ontologies. The applications can be used e.g. for teaching where historic regions have been and how they are related with each other in a partonomy hierarchy. The visualization is made using a rich set of historic maps, modern maps, satellite images, and polygonal boundaries. In addition, the applications can be used for retrieving historical cultural content related to the regions. The relationship is explicated for the user indicating whether the content has been found, used, manufactured, or located in a specific region.

1.6.2   Related Work

Spatio-temporal ontologies for geographic information have been discussed and developed before, especially from a philosophical and foundational viewpoint, and using formal logic approaches [4, 11, 33, 35, 46]. In contrast, the model presented in this chapter is practical, based on simple spatio-temporal relations, and with a focus on the overlap-relation in an ontology time series.

Research on spatio-temporal databases concerns database concepts capturing spatial and temporal aspects of data, including geometry changing over time [32]. Our model is dealing with similar problems but the approach is based on semantic web techniques and ontologies [34], with a focus on dealing with incomplete data, reasoning, data integration, and web applications.

In GIS systems, overlap of physical areas is usually determined by representing the real world in terms of intersecting polygons [37, 43]. However, in application cases like ours, such geometrical modeling may not be feasible because precise geometrical information is not available or it could be difficult to create and computationally difficult to use.

Traditions in ontology versioning [28] and ontology evolution [29] are interested in finding mappings between different ontology versions, doing ontology refinements and other changes in the conceptualization [27, 36], and in reasoning with multi-version ontologies [15]. In ontology mapping research, there have been efforts to do mappings based on probabilistic frameworks [31]. Means for handling inconsistencies between ontology versions [13] have been developed. Methods for modeling temporal RDF have been proposed recently [12]. In contrast to these works, our approach is merely about the evolution of an ontology time series that is due to changes in the underlying domain. Hence it should not be confused with ontology versioning, database evolution, or ontology evolution even if changes are considered in all of these approaches as well. Each temporal member ontology in a time series is a valid, consistent model of the world within the time span it concerns, and may hence be used correctly in e.g. annotation.

Ontology library systems have been proposed for publishing ontologies and providing services for accessing them. Based on reviews on ontology libraries [1,9], the main focus in previously developed systems tends to be in supporting ontology development rather than in providing services for using the ontologies. Although ONKI Ontology Service provides support for the whole ontology life cycle, a major contribution of ONKI is the support for content annotation, information searching and other end-user needs as integrable web widgets and APIs.

Compared to general RDF search engines [6, 8] and ontology servers [7, 30], ONKI Ontology Service is based on an idea of a collection of domain-specific ontology servers providing user interfaces and services suited for ontologies of a given domain. E.g., geographical regions in spatial ontologies can be visualized on a map view.

In information retrieval query expansion techniques have been proposed to solve problems related to the user's ability to represent her information needs in a query adequately [44]. Query expansion can be based on a corpus, e.g. analyzing co-occurrences of terms, or on knowledge models, such as thesauri [45] or ontologies [44]. Methods based on knowledge models are especially useful in cases of short, incomplete query expressions with few terms found in the search index [44, 45].

A spatial query can explicitly contain spatial terms (e.g. Helsinki) and spatial relations (e.g. near), but implicitly it can include even more spatial terms that could be used in query expansion [10], e.g., neighboring places. Spatial terms – i.e. geographical places – do not exist just in space but also in time [21]. Thus, relations between historical places and more contemporary places can be utilized in query expansion. In the ONKI Semantic Query Expansion Service we have used the spatial overlap relation between places to expand the spatial query terms. As query expansion may cause uncontrolled expansion of result sets, thus causing potential loss in the precision of the query [14, 38], the query expansion has been made transparent and controllable to the user.

1.6.3 Future Work

We are currently extending the SAPO ontology to include smaller and older regions. Our RDF repositories already include tens of thousands of places that are being mapped on SAPO and a modern geo-ontology of Finland that consists of hundreds of thousands of places. The idea in a longer perspective is create an ever growing open source RDF repository of historical places in Finland, and link them with international sources, such as TGN and GeoNames.

A further research direction would be to investigate whether the methods and tools presented in this paper could be generalized to other domains, where concepts overcome changes affecting their extensions, properties, or positions in ontological hierarchies and structures.

Acknowledgements This work is part of the National Semantic Web Ontology project in Finland[19] (FinnONTO, 2003–2012), funded mainly by the National Technology and Innovation Agency (Tekes) and a consortium of 38 organizations, and the Cultural Foundation of Finland.

References

1. Ahmad, M.N., Colomb, R.M.: Managing ontologies: a comparative study of ontology servers. In: Proceedings of the eighteenth Conference on Australasian Database (ADC 2007), pp. 13–22. Australian Computer Society, Darlinghurst, Australia, Ballarat, Victoria, Australia (2007)
2. Anonymous: Changes in the names of new municipalities through municipality mergers in japan. In: Technical Papers of the 25th Session of the United Nations Group of Experts on Geographical Names (UNGEGN). Nairobi. Geographical Survey Institute, Ministry of Land, Infrastructure, Transport and Tourism of Japan (2009)
3. Bhogal, J., Macfarlane, A., Smith, P.: A review of ontology based query expansion. Information Processing & Management **43**(4), 866–886 (2007)

---

[19]http://www.seco.tkk.fi/projects/finnonto/.

4. Bittner, T., Donnelly, M., Smith, B.: A spatio-temporal ontology for geographic information integration. Int. J. Geogr. Inf. Sci. **23**(6), 765–798 (2009)
5. Clementine, E., Felice, P.D., van Oosterom, P.: A small set of formal topological relationships suitable for end-user interaction. In: A. Abel, B.C. Ooi (eds.) Advances in databases, pp. 277–295. Springer–Verlag (1993)
6. d'Aquin, M., Baldassarre, C., Gridinoc, L., Sabou, M., Angeletou, S., Motta, E.: Watson: Supporting next generation semantic web applications. In: Proceedings of IADIS International Conference on WWW/Internet. Vila Real, Portugal (2007)
7. d'Aquin, M., Lewen, H.: Cupboard - a place to expose your ontologies to applications and the community. In: Proceedings of the ESWC 2009, Heraklion, Greece, pp. 913–918. Springer–Verlag (2009)
8. Ding, L., Finin, T., Joshi, A., Pan, R., Cost, R.S., Peng, Y., Reddivari, P., Doshi, V., Sachs, J.: Swoogle: a search and metadata engine for the semantic web. In: Proceedings of the ACM International Conference on Information and Knowledge Management, pp. 652–659. ACM (2004)
9. Ding, Y., Fensel, D.: Ontology library systems: The key to successful ontology reuse. In: Proceedings of SWWS'01, The first Semantic Web Working Symposium, Stanford University, USA, pp. 93–112 (2001)
10. Fu, G., Jones, C.B., Abdelmoty, A.I.: Ontology-based spatial query expansion in information retrieval. In: On The Move to Meaningful Internet Systems: ODBASE 2005, Agia Napa, Cyprus, pp. 1466–1482. Springer–Verlag (2005)
11. Grenon, P., Smith, B.: SNAP and SPAN: Prolegomenon to geodynamic ontology. Spatial Cognition and Computation **4**(1), 69–104 (2004)
12. Gutierrez, C., Hurtado, C., Vaisman, A.: Temporal RDF. In: 2nd European Semantic Web Conference (ESWC2005), pp. 93–107. Heraklion, Greece (2005)
13. Haase, P., van Harmelen, F., Huang, Z., Stuckenschmidt, H., Sure, Y.: A framework for handling inconsistency in changing ontologies. In: International Semantic Web Conference, pp. 563–577. Springer (Galway, Ireland, November 6-10,2005)
14. Hollink, L., Schreiber, G., Wielinga, B.: Patterns of semantic relations to improve image content search. Journal of Web Semantics **5**(3), 195–203 (2007)
15. Huang, Z., Stuckenschmidt, H.: Reasoning with multi-version ontologies: A temporal logic approach. In: International Semantic Web Conference, pp. 398–412 (2005)
16. Hyvönen, E.: Semantic Portals for Cultural Heritage. In: Staab and Studer [34] (2009)
17. Hyvönen, E., Lindroos, R., Kauppinen, T., Henriksson, R.: An ontology service for geographical content. In: Poster Proceedings of the International Semantic Web Conference (ISWC 2007), Busan, Korea (2007)
18. Hyvönen, E., Mäkelä, E., Kauppinen, T., Alm, O., Kurki, J., Ruotsalo, T., Seppälä, K., Takala, J., Puputti, K., Kuittinen, H., Viljanen, K., Tuominen, J., Palonen, T., Frosterus, M., Sinkkilä, R., Paakkarinen, P., Laitio, J., Nyberg, K.: CultureSampo—Finnish culture on the Semantic Web 2.0. Thematic perspectives for the end-user. In: Museums and the Web 2009 Proceedings. Archives & Museum Informatics, Toronto (2009). http://www.archimuse.com/mw2009/papers/hyvonen/hyvonen.html
19. Hyvönen, E., Viljanen, K., Tuominen, J., Seppälä, K.: Building a national semantic web ontology and ontology service infrastructure—the FinnONTO approach. In: Proceedings of the ESWC 2008, Tenerife, Spain. Springer–Verlag (2008)
20. Järvelin, K., Kekäläinen, J., Niemi, T.: ExpansionTool: Concept-based query expansion and construction. Information Retrieval **4**(3/4), 231–255 (2001)
21. Jones, C.B., Abdelmoty, A.I., Fu, G.: Maintaining ontologies for geographical information retrieval on the web. In: On the Move to Meaningful Internet Systems: ODBASE 2003, Sicily, Italy, pp. 934–951. Springer–Verlag (2003)
22. Kauppinen, T., Deichstetter, C., Hyvönen, E.: Temp-o-map: Ontology-based search and visualization of spatio-temporal maps. In: Demo track at the European Semantic Web Conference ESWC 2007, Innsbruck, Austria. Springer–Verlag (2007)

23. Kauppinen, T., Hyvönen, E.: Modeling and Reasoning about Changes in Ontology Time Series, pp. 319–338. In: Kishore et al. [26] (2007)
24. Kauppinen, T., Paakkarinen, P., Eetu Mäkelä Heini Kuittinen, J.V., Hyvönen, E.: Geospatiotemporal semantic web for cultural heritage. In: Digital Culture and E-Tourism: Technologies, Applications and Management Approaches. IGI Global, Hershey, Pennsylvania (2011)
25. Kauppinen, T., Väätäinen, J., Hyvönen, E.: Creating and using geospatial ontology time series in a semantic cultural heritage portal. In: Proceedings of the ESWC 2008, Tenerife, Spain, pp. 110–123. Springer–Verlag (2008)
26. Kishore, R., Ramesh, R., Sharman, R. (eds.): Ontologies: A Handbook of Principles, Concepts and Applications in Information Systems. Springer–Verlag (2007)
27. Klein, M.: Change management for distributed ontologies. Ph.D. thesis, Vrije Universiteit Amsterdam (2004)
28. Klein, M., Fensel, D.: Ontology versioning on the Semantic Web. In: Proceedings of the International Semantic Web Working Symposium (SWWS), pp. 75–91. Stanford University, California, USA (2001)
29. Noy, N., Klein, M.: Ontology evolution: Not the same as schema evolution. Knowledge and Information Systems 5 (2003)
30. Noy, N.F., Shah, N.H., Whetzel, P.L., Dai, B., Dorf, M., Griffith, N., Jonquet, C., Rubin, D.L., Storey, M.A., Chute, C.G., Musen, M.A.: BioPortal: ontologies and integrated data resources at the click of a mouse. Nucleic Acids Research 37(Web Server issue), 170–173 (2009)
31. Pan, R., Ding, Z., Yu, Y., Peng, Y.: A bayesian network approach to ontology mapping. In: International Semantic Web Conference 2005, pp. 563–577. Springer–Verlag (Galway, Ireland, November 6-10,2005)
32. Pelekis, N., Theodoulidis, B., Kopanakis, I., Theodoridis, Y.: Literature review of spatiotemporal database models. The Knowledge Engineering Review 19, 235–274 (2005)
33. Sider, T.: Four-Dimensionalism. An Ontology of Persistence and Time. Clarendon Press, Oxford (2001)
34. Staab, S., Studer, R. (eds.): Handbook on ontologies (2nd Edition). Springer–Verlag (2009)
35. Stell, J.G., West, M.: A 4-dimensionalist mereotopology. In: A. Varzi, L. Vieu (eds.) Formal Ontology in Information Systems, pp. 261–272. IOS Press (2004)
36. Stojanovic, L.: Methods and tools for ontology evolution. Ph.D. thesis, University of Karlsruhe, Germany (2004)
37. Stuckenschmidt, H., Harmelen, F.V.: Information Sharing on the Semantic Web. Springer–Verlag (2004)
38. Tudhope, D., Alani, H., Jones, C.: Augmenting thesaurus relationships: Possibilities for retrieval. Journal of Digital Information 1(8) (2001)
39. Tuominen, J., Frosterus, M., Viljanen, K., Hyvönen, E.: ONKI SKOS server for publishing and utilizing SKOS vocabularies and ontologies as services. In: Proceedings of the ESWC 2009, Heraklion, Greece, pp. 768–780. Springer–Verlag (2009)
40. Tuominen, J., Kauppinen, T., Viljanen, K., Hyvönen, E.: Ontology-based query expansion widget for information retrieval. In: Proceedings of the 5th Workshop on Scripting and Development for the Semantic Web (SFSW 2009), 6th European Semantic Web Conference (ESWC 2009), vol. 449. CEUR Wokshop Proceedings, http://ceur-ws.org/ (2009)
41. Viljanen, K., Tuominen, J., Hyvönen, E.: Ontology libraries for production use: The Finnish ontology library service ONKI. In: Proceedings of the ESWC 2009, Heraklion, Greece, pp. 781–795. Springer–Verlag (2009)
42. Viljanen, K., Tuominen, J., Salonoja, M., Hyvönen, E.: Linked open ontology services. In: Workshop on Ontology Repositories and Editors for the Semantic Web (ORES 2010), the Extended Semantic Web Conference ESWC 2010, vol. 596. CEUR Workshop Proceedings, http://ceur-ws.org/ (2010)
43. Visser, U.: Intelligent information integration for the Semantic Web. Springer–Verlag (2004)

44. Voorhees, E.M.: Query expansion using lexical-semantic relations. In: Proceedings of the 17th annual international ACM SIGIR conference on Research and development in information retrieval, Dublin, Ireland, pp. 61–69. ACM (1994)
45. Wang, Y.C., Vandendorpe, J., Evens, M.: Relational thesauri in information retrieval. Journal of the American Society for Information Science **36**(1), 15–27 (1985)
46. West, M.: Replaceable parts: A four dimensional analysis. In: COSIT-03 - Workshop on fundamental issues in spatial and geographic ontologies. Ittingen, Switzerland (2003)

Chapter 2
Semantic Referencing of Geosensor Data and Volunteered Geographic Information

Simon Scheider, Carsten Keßler, Jens Ortmann, Anusuriya Devaraju,
Johannes Trame, Tomi Kauppinen, and Werner Kuhn

Abstract Georeferencing and semantic annotations improve the findability of geoinformation because they exploit relationships to existing data and hence facilitate queries. Unlike georeferencing, which grounds location information in reference points on the earth's surface, semantic annotations often lack relations to entities of shared experience. We suggest an approach to semantically reference geoinformation based on underlying observations, relating data to observable entities and actions. After discussing an ontology for an observer's domain of experience, we demonstrate our approach through two use cases. First, we show how to distinguish geosensors based on observed properties and abstracting from technical implementations. Second, we show how to complement annotations of volunteered geographic information with observed affordances.

2.1 Introduction and Motivation

Observations are the principal source of geographic information. Humans share senses[1] and perceptual capabilities [1] that enable them to observe their environment, and thereby obtain geographic information. For example, vision works essentially the same way for all humans. Additionally, humans can easily understand and reproduce observations made by others, because they can understand intentions and join their attention in a scene [2]. If someone tells you that Main Street is closed due to construction works, you can easily understand what was observed without observing it yourself. Some of the authors of this chapter have previously suggested

---

[1] With few exceptions, such as disabilities, that do not affect the general case.

S. Scheider (✉)
Institute for Geoinformatics, University of Münster, Münster, Germany
e-mail: simon.scheider@uni-muenster.de

N. Ashish and A.P. Sheth (eds.), *Geospatial Semantics and the Semantic Web:*
Foundations, Algorithms, and Applications, Semantic Web and Beyond 12,
DOI 10.1007/978-1-4419-9446-2_2, © Springer Science+Business Media, LLC 2011

to use perceptual capabilities as common ground to describe geoinformation [3, 4]. In this chapter, we demonstrate how to account for the semantics of geoinformation based on underlying observations. Our approach is general enough to account for observations obtained from technical sensors (such as a thermometer measuring temperature) as well as human observations (e.g., observing the presence of a construction site on Main Street).

Dealing with the semantics of geoinformation in terms of observable properties (such as temperature, precipitation rate, or traversability of a road), we face the problem of finding an appropriate level of description. This problem is our main focus. It has two aspects. On the one hand, there is a plethora of different sensing procedures for the same property that lead to equivalent results. Hence, their differences are irrelevant for the meaning of the obtained geoinformation. For example, precipitation rate can be measured by a tipping-bucket or a standard rain gauge. However, the meaning of a value of five liters of rainfall in the last 24 h is independent of the concrete form of the sensor. Therefore, this description is too detailed to describe the property. The problem of having an unnecessarily detailed description of the semantics of geoinformation is called the *abstraction problem*.

On the other hand, we have a *grounding problem* [5]. This problem occurs when it is not clear what kind of observation certain information refers to. One of the most dramatic examples of this was the *moon-alarm* bringing the world to the brink of a nuclear war: On 1st October, 1960, the brand new Ballistic Missile Early Warning System of the United States Air Force took radar signals reflected by the moon for Russian missiles. Luckily, human reason prevented the nuclear "counter"-attack (cf. [6]). Less dramatic, but more frequently, the grounding problem occurs if measurements are only described by SI units.[2] A velocity value of 2 given in meters per second just tells us that there is something moving, but we cannot even tell whether it is a car on the road, gravel on a slope, water in a riverbed, or anything else. Volunteered Geographic Information (VGI) [7], which is publicly available on the Web, faces a similar problem. The most prominent collection of VGI is the OpenStreetMap (OSM) project,[3] where users have the opportunity to describe map features via tags. However, the tags that are used to describe points of interest (POI) often do not make clear what the *interest* in a specific point is. That is, they do not provide sufficient information about what is afforded by the POI: The tag `cafe` is used to describe *coffee shops* in New York as well as *Kaffeehäuser* in Vienna. If a user wants to have a beer, a place tagged `cafe` in Vienna is perfectly suitable, whereas a coffee shop tagged `cafe` in New York is not. Here the appropriate level of abstraction would rather be on the level of observed functional properties, like `drinkBeer` or `drinkCoffee`.

State-of-the-art approaches to modeling the semantics of geoinformation do not seem to provide an appropriate level of abstraction. Current top-level ontologies, e.g. the Descriptive Ontology of Linguistic and Cognitive Engineering (DOLCE) [8] or the Basic Formal Ontology (BFO) [9], clarify ontological commitments, but

---

[2]Le Système international d'unités, see http://www.bipm.org/en/si/.

[3]See http://www.openstreetmap.org/.

abstract from observation procedures. Therefore, they provide only a partial solution to our problem. Similarly, VGI often relies on user-defined domain specific tags, which lack an unambiguous interpretation in terms of reproducible observations. Current metadata standards, like the Observations and Measurements specification of the Open Geospatial Consortium[4] [10], describe geosensor data at the level of information objects, not of observed properties or objects [11].

The idea to use reproducible observations to describe the semantics of geoinformation is not new. Geodesists are routinely grounding coordinates in reproducible measurements of distances and directions. The reference points and parameters for these measurements define *geodetic datums*. We follow here Kuhn's [12] generalization from spatial to *semantic* reference systems to describe the semantics of arbitrary geographic information (not just locations). To construct semantic reference systems in practice, we have suggested conventional semantic datums in terms of repeatable observation procedures [13].

In this chapter we discuss a set of *perceptual types* for describing observations underlying geographic information. Perceptual types are types of entities in an observer's domain of sensory experience. We argue that these comprise *perceptual Gestalts*[5] such as observed bodies, media, surfaces, actions and properties. Grounding our ontology in perceptual types has the advantage that these provide a direct link between the world experienced by an observer and a top-level ontology. We will show that perceptual types neatly fit into DOLCE. Observations also ensure semantic interoperability [14] in the sense that they are easily reproducible by different observers. Additionally, semantic referencing of geoinformation based on observations and perceptual types provides an *appropriate level of abstraction* for annotating and querying geoinformation. Observations and perceptual types allow us to abstract from technical measurement procedures, while differentiating among observed properties beyond SI units and among perceivable functional properties.

Modeling an ontology should be distinguished from implementing it in an encoding language like OWL [15]. To maintain sufficient expressivity when modeling our ontological theory, we use a typed first-order logic with functions. An essential subset of our model can be encoded in RDF,[6] allowing to link data to perceived entities and to publish the results on the Semantic Web. Note that this subset in RDF does not exceed the boundaries of existing semantic web technology. Therefore, the paper also demonstrates what can be expressed by current semantic web standards.

The remainder of this chapter is structured as follows. In Sect. 2.2 we review background notions to describe human experience. These include affordances, media, and bodies. In Sect. 2.3 we introduce perceptual types and operations in a functional first-order style and align them with categories of the top-level ontology DOLCE. In Sects. 2.4 and 2.5 these types are used to describe two scenarios, one concerned with technical sensors and one with VGI. We conclude the chapter in Sect. 2.6 with a review and outlook.

---

[4]See http://www.opengeospatial.org/.

[5]For more information on Gestalt perception see [1].

[6]See http://www.w3.org/RDF/.

2.2 Background

In this section, we make some justifiable claims about what kinds of perceptual capabilities humans have in order to observe geoinformation. Technical sensors are extensions of human senses. They need to be designed, built, calibrated, maintained and interpreted by humans. This allows us, in principle, to trace back the observation of technical sensors to human perceptual capabilities. The example of the moon alarm mentioned in Sect. 2.1 highlights the necessity of human observers as interpreters and controllers of technical sensors.

2.2.1 Perceiving the Meaningful Environment

The ecological psychologist Gibson [16] suggested an informal ontology of elements of the environment that are accessible to human perception and action, called the *meaningful environment*. The three top-level categories of *meaningful things* [16, p. 33] in this environment are *substances*, *media* and *surfaces*.

A medium affords moving through it as well as seeing, smelling and breathing and bears the perceivable vertical axis of gravity (for vertical orientation). According to Gibson, the medium for terrestrial animals is the air. Gibson thought of a medium as something established in terms of *affordances*, i.e. action potentials in the environment. For example, he distinguished liquid media (water) and gaseous ones (air) by what actions they afford to the animal [16, Chapter 2]. We have suggested [13] that there may be different kinds of media according to what kind of action they offer to a human being. In this chapter, we restrict our understanding of a medium based on locomotion and action affordances. This view will be explained in the next section.

Surfaces are the boundaries of all meaningful things humans can distinguish by perception. This means they are opaque to a certain extent and bound an *illuminated medium*, i.e., a medium for seeing. Surfaces have *surface qualities*, for example a *texture* (including color), and are often resistant to pressure.

Substances are things in the environment that are impenetrable to motion (i.e., are solid) and illumination (i.e., are opaque). Detachable substances are called *objects*, which have further properties, e.g. a shape and a weight. Moreover, substances enable actions: they support movements (the ground), they enclose something as hollow objects, or they allow to be thrown as detached objects.

One of Gibson's central insights was that the elements of the meaningful environment are inter-subjectively available to human observers in their domain of experience. However, if one does not assume that observers have direct access to external reality [17], this can only mean that they have analogous criteria or capabilities for *identifying and distinguishing* these things. We have suggested [3] that some of these meaningful things could be viewed as results of mental constructions [18] based on preconceptually available Gestalt mechanisms [1], for example identifying and tracking bodies and their surfaces [19]. Complex qualities of bodies can be constructed by performing *perceptual operations* on their surface layout, e.g. by observing their lengths or depths [13]. Movements and other events can be individuated by following these bodies with attention [3]. Media can be individuated based on the affordances they offer an observer [13,20]. For example the affordance of locomotion identifies the medium that allows you to travel. This can be just the free space of your office, if the door is closed, or extend several kilometers throughout the landscape if you are hiking outside.

Individuation requires criteria of *unity* (i.e., for constructing integral wholes as maximal self-connected sums) and *identity* (i.e., allowing to track entities and distinguish them from each other) [21]. In this chapter, we will assume that there are criteria of individuation available for all perceptual types mentioned in Sect. 2.3.2, without discussing how the resulting entities can be constructed in experience.[7] We furthermore presume the existence of reference systems for complex qualities (like velocity, volume or weight).

2.2.2 Perceived Affordance: A Simulative Account

Affordance is one of the key concepts in ecological psychology. Affordances capture the functional aspect of objects in an observer's environment as well as an observer's opportunities for actions [22]. As Gibson puts it:

> "The *affordances* of the environment are what it *offers* the animal, what it *provides* or *furnishes*, whether for good or ill. [...] I mean by it something that refers to both the environment and the animal [...]." [16, p. 127, emphasis in original]

An observer in this view is not only perceiving but also (potentially) acting. Gibson's own examples of affordances include *action affordances* like climb-ability (walls), catch-ability (balls), eat-ability, mail-ability (postbox), but also so-called *happening affordances* like getting burned (by fire) or falling off (a cliff) (compare [23]).

Viewing affordances as properties of things in the environment [24] seems problematic, because they are also constituted by properties of a particular agent: Stairs are climbable only with respect to an agent's leg length (cf. Warren's experiments [25]). Treating affordances as combined qualities of environments and actors (as proposed in [26]), which seems to work in the staircase example (by relating leg length and riser height), is also problematic. Take, for example the *traversability* of a road. A road is traversable with respect to the velocity of an agent's crossing and the velocity of cars. But traversability is not a combination of a property of the agent with a property of the environment. Rather, it is the interplay of objects which is not

---

[7] See [13, 20] for examples how this might be done.

reducible to any combination of properties (cf. [27]). We follow Scarantino [23] in that affordances always involve an observer's reaction. We conceive of them as *perceivable potential events*.

But how are potential events perceived? One possible explanation is that perceived affordances are the result of *perceptual simulations*. These were proposed by Barsalou [28] in order to state that human perception and cognition are closely interlinked on the basis of perceptual simulators. They allow humans to imagine and reconstruct formerly perceived sensori-motor patterns of objects, e.g. cars, in new situations, and to reason with them. We suggest to apply this idea to affordances, saying that if pedestrians perceive the affordance of crossing a road, they do so by successfully simulating a crossing event in a given perceived scene. Perceived affordances can be "acted on", i.e. they are a necessary input to human actions, as proposed by Ortmann and Kuhn [29]. Similarly, when placing a rain gauge, we simulate potential raining events in order to set it up in a medium for rain, e.g. in our garden instead of our living room.

Many affordances have a social aspect, in the sense that they involve the interpretation of signs. A prominent example for a so-called *social affordance* [30] is a postbox that affords sending letters. The postbox physically only affords dropping letters (or other similarly shaped objects) through a slot. However, in the social environment that uses the appearance of boxes as conventional signs (blue in the USA, red in the UK, yellow in Germany), this box affords sending letters if the letters are properly labeled and postpaid. Since a simulative account of affordances does not exclude cognition of signs, social affordances are compatible with our approach.

2.2.3 Structuring Perceptual Types with DOLCE

We use the DOLCE[8] [8] as a top-level (or foundational) ontology for structuring the perceptual types proposed in Sect. 2.3. DOLCE rests on four foundational categories: Endurants, Perdurants, Qualities and Abstracts. *Endurants* are things that are fully present at any moment, but can change over time. Examples of Endurants are all physical objects, such as streets, cars, trees, buildings, as well as amounts of matter (e.g., water, air, sand or concrete), but also features like a crack in a street or a hole in a wall. *Perdurants* are entities that are not fully present at any time. Perdurants occupy a time span. For example, a football match, a thunderstorm or a lunch break all last for a certain time. Endurants typically participate in Perdurants. You, your colleagues and your lunch are participating in your lunch break. Amounts of rain, amounts of air and the city on the ground participate in a thunderstorm. *Qualities* inhere in other entities and are similar to common sense properties.

---

[8]http://www.loa-cnr.it/DOLCE.html.

Examples are the height quality of a step, the velocity of a current or the duration of a thunderstorm. In general, all physical endurants have a spatial quality and all perdurants have a temporal quality.

DOLCE has been applied to geospatial ontologies, among others, to describe geographic entities in geology [31], to provide a foundational model of geographic entities [32], to ground the SWEET Ontology [33], as well as to establish semantic reference systems for observations and measurements [34] and to ground an observation ontology [29, 35].

DOLCE has been proposed for developing sound ontologies. For an information category to be *ontologically sound*, identity criteria are required [36]. In our view, the application and combination of perceptual Gestalt operations establishes criteria of identity for environmental entities [3]. It is therefore not surprising that many ontologically sound top-level categories (called *sortals* in [21]), such as the ones of DOLCE [8], can be aligned with this lower perceptual level. This will be demonstrated in Sect. 2.3.2.

2.3 Grounding Geospatial Data in Perceptual Types

In the following we introduce and explain basic perceptual types that we use for grounding information in the scenarios of the subsequent sections. Based on the discussion in the last section, we suggest that for all types there are individuation criteria available to human observers, enabling them to track and distinguish instances of a type. Consequently, observations are described from the perspective of a human observer.

2.3.1 Notation for Perceptual Operations and Types

We use a typed first-order logic for describing an observer's domain of experience, in which types $T_i \in T$ are used in type assignments of the form:

$$f : T_1 \times \ldots \times T_r \mapsto T_{r+1} \quad \text{for a function } f,$$

$$P : T_1 \times \ldots \times T_r \quad \text{for predicates } P,$$

$$c : T_i \quad \text{for constants } c \text{ or variables.}$$

Types are introduced with the prescript type. Type as well as predicate symbols start with uppercase letters, constants and variables are lowercase. Unary type symbols are used interchangeably with unary predicates, for example $c : T_i$ means $T_i(c)$ where T_i is used as a predicate symbol. Basic types correspond to primitive predicates. We use $\lor$ and $\land$ to construct dis- and conjunctive unary types. N-ary types can be constructed using $\times$ (product) and $\mapsto$ (function) type constructors.

Perceptual operations are expressed as functions that are applied by an observer to entities and produce entities in his or her domain of experience. They may look like this: $Op : T_i \mapsto T_j$, where T_i is the input type, and T_j is the type of the observation result. If the operator has more than one input of the same type, we may also write $Op : T_i^* \mapsto T_j$ to denote this. Any predicate may also be written as a function that maps to entities of boolean type, e.g. $P : T_1 \times \ldots \times T_r \mapsto Bool$. We do not intend to list perceptual operations exhaustively, because for every domain, we may have special subtypes of them. Therefore, most of the operations are given as function schemas. Signatures and explanations of these schemas will be given in the text. We introduce perceptual types with a minimal formal apparatus for the sake of demonstration.

2.3.2   Unary Perceptual Types and Their Hierarchy

The entities which can be distinguished in experience come with their categories or unary types. These types can be arranged in a subsumption hierarchy (see Fig. 2.1) aligned with some of DOLCE's top-level categories.[9]

The most important types are perceivable *bodies* (type *Body*) as self-connected, solid, movable objects. We distinguish type *Animate* (e.g. human) bodies and type *Inanimate* bodies.[10] The empty space that contains these bodies, which is the medium of their movements and actions, is conceived as a maximal part of the environment that affords bodies to move and act in them. We call such an entity a *medium* (type *Medium*). We also allow media to afford events for inanimate bodies, e.g. a cliff to afford falling rocks.

Media and bodies have a criterion of unity and are rigid types [36], whose instances can be identified in time. We consider therefore a *physical object* (type *PhysicalObject*), in extension of DOLCE, as being either a medium or a body (see Fig. 2.1). One of the perceived bodies is the body of the observer, and one of the media is the one surrounding him or her allowing to move or act. As this medium is identified via an affordance, it moves as soon as the perceived affordance changes its location. For example, if the door is being closed, the medium suddenly reduces to the room.

We furthermore assume that there is a range of independent *subtypes of media*. These depend on the type of object and the type of motion or action the medium affords to the object. For example, a water body is part of a general motion medium for inanimate bodies, including the air but excluding the ground. For instance, a stone can fall through water and air, but not through the ground. The *water unit*

---

[9]However, we sometimes divert from strictly following DOLCE and explain this in the text.

[10]For simplicity reasons, we do not consider animate (agentive) objects as constituted from inanimate (non-agentive) ones, as DOLCE does [8], but see them as subcategories of bodies distinguished according to perceived intentionality. Intentionality is thus constitutive for perceived actions, and actions for animates.

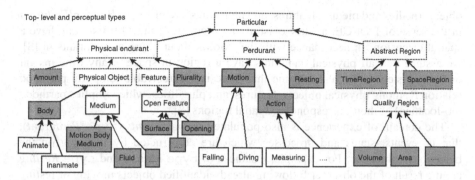

Fig. 2.1 Subsumption hierarchy of perceptual and top-level types. *Basic types* with identity criteria are *highlighted* in *orange*. The *dotted boxes* correspond to top-level categories of DOLCE

part of this medium,[11] on the other hand, is by itself a medium for fish or divers. Likewise, the upper part of the ground is a medium for a mole, and a snowpack is a medium for rescue dogs or snow stakes. Part of the reason why media afford a certain type of motion is their low physical resistance to movement and tensile stress, i.e., their *viscosity*. Therefore, we suggest that it is the affordance of a certain kind of "forceful" motion allowing observers to categorize media of type *Fluid*. Fluids can be recognized depending on their resistance to stress applied by an object moving through it. The concept of a medium can also be used in its normal context of human actions. We may perceive certain media in the environment based on *social affordances*. These allow people to act based on the interpretation of social conventions and signs, e.g. to drive on a certain marked surface identified as a lane of a road [20].

We assume that the observer's domain of experience also contains distinguishable parts of bodies and media on which to focus one's attention. Some of these parts are what DOLCE calls *features* (type *Feature*). Features have their own criterion of unity, but depend on another physical object, their "host". While a feature needs a host, it does not need to be part of it. Perceivable features of a cup, for example, are its handle but also its opening. The opening of a cup would not exist without it, but is not a part of the cup. A feature of a building is the opening of its entrance. Another important feature is the *surface* of an object (type *Surface*), which, in the sense of Gibson [16], can be conceived as the border of a surrounding illuminated medium that affords seeing. Surfaces are themselves hosts for surface qualities like texture and color. In accordance with DOLCE, the experiential domain must also contain arbitrary sums: for example the sum of cars driving past a house and the sum of their wheels. These dependent entities are of type *Plurality*. All these entities, physical

---

[11] In the remainder, we use the term "water unit" or "unit of water" for any fluid medium consisting of water, regardless of its size. The term "water body" is commonly reserved for large water units that are physiographical features.

objects (bodies and media), features, and pluralities, are of type *PhysicalEndurant* in the sense of DOLCE, as they exist at any moment of their lifespan and have a spatial extension. In accordance with the axioms about physical endurants in [8], we assume that all physical endurants have a region in space at any time instant of their existence. We also assume that there is a perceivable temporal parthood relation *P* among physical objects, features, and pluralities, which implies parthood (co-location) of their corresponding spatial regions [8].

The domain of experience is also populated by *perdurants* (type *Perdurant*), that is, events, states and processes which are constructed by focusing on the behavior of objects.[12] For example, instances of type *Motion* and type *Resting* are the result of the observer following already identified objects moving or resting in time and space. In accordance with DOLCE, we assume that all perdurants occupy a region in time. Note that determining whether some object moves or not is always observer-relative. A resting therefore simply denotes objects at rest from the perspective of some spatial reference system. If the observer imputes *an intention* to the object involved in a perceived event, this event is of type *Action*.

We assume that the observer can determine the *location* of a physical endurant in a spatial reference system, and the *time interval* of a perdurant in a temporal reference system. The observer has also reference systems for certain other quality types, such as the volume of an endurant, or the area of a flat object, or the color of its surface patch. Their values are of type *AbstractRegion* and are part of a quality space [8, 11].

There is a last important category of DOLCE which we conceive as a perceptual type, namely the type *Amount*.[13] This notion can be applied to individual portions of matter contained in some object, e.g. the portion of clay of which a statue is made, or the portion of water flowing through some river into the sea. The authors of DOLCE did not consider unity or identity criteria for amounts [8]. Moreover, amounts do not seem to be directly perceivable, since we cannot identify portions of matter as such. One can see this by the fact that we perceive snow packs and lakes in terms of their surfaces, while the matter they contain constantly keeps being exchanged by melting, evaporation or discharge without notice. Nonetheless, we follow Guizzardi [37] and assume that amounts can be *identified indirectly* via objects with a unity criterion (e.g. a certain water unit with perceivable surfaces) and their perceived histories. This assumption is vital in order to deal with the observation of river flow, as we will see in Sect. 2.4.4. We will discuss perceptual operators for amounts in Sect. 2.3.4.

All these entities are related as shown in the subsumption hierarchy depicted in Fig. 2.1.

---

[12]We do not distinguish here between state, process and event perdurants as in DOLCE, also because we are unsure of whether our perceptual types should be regarded as cumulative or not, compare [8].

[13]Known as "amount of matter" in DOLCE.

2.3.3   Basic Types and Perceptual Operations

Our idea is that perception is the key to distinguish, i.e. to identify, entities. In some sense this means that it accounts for how experiential entities come into being.

This aspect is reflected in our type hierarchy, because some types come with their own identity criterion. These types are basic in that they *carry their own criterion of identity (IC)*, while their subtypes only inherit these criteria, and their supertypes are just aggregations (disjunctive types) without any such criterion. For example, we can distinguish one person from another on the level of their bodies ("this body is distinct from that one"), but not on the level of *Animate* (subtype, inherited IC) or *PhysicalObject* (supertype). These types may be called *basic types*[14] or *true sortals* [36] and are highlighted in the hierarchy of Fig. 2.1. In the same sense, the concrete kind of perceived affordance, differentiated by the type of body and the type of motion or action involved, gives an identity criterion for media subtypes (see Fig. 2.1). Therefore the basic types of media are given on the level of the subtypes of the type medium. For example, a *MotionBodyMedium* has a different identity criterion than a *FallingBodyMedium*, but the type *Medium* itself does not have any IC. Since ICs define identity (=) between entities of a basic type, we can assume that *all basic types are mutually exclusive*, because there cannot be identical entities across those types. This applies also to media, because even though every *Falling* is an instance of *Motion*, a *FallingBodyMedium* never is a *MotionBodyMedium*. However, since a place that affords falling also affords moving, and media are constructed out of such affordances, a *FallingBodyMedium always implies* a larger *MotionBodyMedium* of which it is a part.[15]

We will not discuss how basic types can be constructed from perceptual mechanisms, i.e., how these identity criteria are actually established. But we will introduce *perceptual operations* as functions in order to highlight how they depend on each other. The formal properties of these functions will be discussed now.

A perceptual operator function is not necessarily *total*, so for some input it may produce errors.[16] This corresponds to the fact that not every observable input of an observation process gives rise to a certain kind of observation result. For example, not every observable object has a length or is involved in a perceived movement or action.

In some cases, the operators express *existential dependence* of outputs on inputs. Another way of stating existential dependence (compare [8]) is to say they are

---

[14]In the spirit of basic-level categories [38]. But this notion also stands for non-constructed types. Since our basic types correspond to primitive predicates in our theory, both senses are applicable here.

[15]This can be inferred formally if media are defined as maximal integral wholes (i.e., connected sums) of places affording motions/actions [20]. In this case the place must be part of two integral wholes.

[16]We assume that there is one standard error element produced by every function in that case.

surjective, so if there exists an instance of their output type, then there must also exist a corresponding instance of the input type that has generated the output. For example, a process only exists insofar as its participating objects exist, and a feature only exists insofar its host exists. Furthermore, a feature only has one particular host, and a process only has a fixed set of objects that generate it (*injectivity*). Together these properties are called *bijectivity* and allow the observer to distinguish the generated entities via the inputs to the perceptual operator. We assume bijectivity only for those operators Op whose output O is a subtype of a basic type of *Perdurant* or *Feature*. The individuation of bodies, media, qualities and amounts of matter is more complicated and out of the scope of this chapter.

2.3.4 Some Basic Examples of Perceptual Operations

In the following, perceptual operations are introduced using type signature schemas. In these schemas, $[[T]Type]$ is a meta-variable ranging over all subtypes of a type T. For example, $[MotionType]$ ranges over subtypes of *Motion*. Concrete signatures and axioms are obtained by substituting these subtypes.

Operation for perceiving parthood among (non-amount) endurants. This is an operation that allows the observer to relate endurants which are not amounts in the perceived environment, i.e. objects, features and pluralities. It corresponds to DOLCE's temporal parthood [8]. P is therefore a mereological operator which is *not extensional*,[17] as two different objects, e.g. two media, may have the same parts and may be co-located, and objects may change parts. P also implies spatial inclusion among the object's locations [8].

$$(\text{temp. parthood})P : NonAmountE \times NonAmountE \times TimeRegion \mapsto Bool \quad (2.1)$$

The notion *NonAmountE* stands for the type *PhysicalObject* $\vee$ *Feature* $\vee$ *Plurality*. If we omit the temporal parameter of this predicate, we simply mean that parthood is observed continuously, i.e. for every possible time interval.

Operations for perceiving perdurants. These operations take one or several endurants and a time interval and produce a movement or action perdurant in which the endurants participate. They are similar to DOLCE's *participation* relation [8], but allow to distinguish intentional (actions) from non-intentional events. They may also involve many endurant instances as arguments. In order to express that the observer's attention follows objects for an interval in time and identifies one of their movements, we use the operator *move*. We distinguish kinds of movements

---

[17]See Sect. 3.3 in [39].

by inserting the movement type into the operation name. For example, we assume the movement type *Diving*, and so we can express observed divings by the operator *moveDiving*:

$$move[MotionType] : Endurant^* \times TimeRegion \mapsto [MotionType] \qquad (2.2)$$

We follow animate bodies producing actions by the operator *do*. Actions are not necessarily associated with movements. We assume that the first parameter of *do* is the actor's body, and that there may be further optional endurants involved in that action:

$$do[ActionType] : Animate \times Endurant^* \times TimeRegion \mapsto [ActionType] \qquad (2.3)$$

For example, we will make use of the action type *Measuring*, and thus a particular measuring action can be expressed by the operator *doMeasuring*. Another operation called *rest* observes some endurant at rest:

$$rest : Endurant \times TimeRegion \mapsto Resting \qquad (2.4)$$

As discussed in Sect. 2.2.2, there are cases in which observers do not watch body movements or actions being performed explicitly (or watch others performing it), they only simulate them in the perceived environment in order to find out whether they are afforded. For example, in order to guess whether it is possible to climb over a fence, or whether a stone would fall into a well. The output of these simulations are also action perdurants. We assume that every type of observable action or movement can be simulated. For the sake of simplicity, we do not distinguish here whether actions and movements are simulated or actually performed.

Operations for measuring time intervals of perdurants. We measure time in terms of temporal reference systems. These scales are based on calibrated clocks and calendars. The observation process needs a perdurant as input, whose temporal extent is measured. We call this operation *time*:

$$time : Perdurant \mapsto TimeRegion \qquad (2.5)$$

Another operation allows for measuring durations of time intervals. This is done by subtracting the beginning from the end of a temporal region, which produces just another time interval which denotes the duration (compare also Fig. 2.3).

$$duration : TimeRegion \mapsto TimeRegion \qquad (2.6)$$

Operations for measuring locations and other qualities of endurants. The first operation localizes the spatial region that corresponds to a certain physical endurant in a spatial reference system at a certain time. The other operations are examples for basic observations of qualities. All of them correspond to DOLCE's physical qualities [8]. They produce an output region whose temporal resolution depends on the input time interval. For example, observing the location of a moving object for an interval results in an extended region that encompasses this movement.

$$location : Endurant \times TimeRegion \mapsto SpaceRegion \qquad (2.7)$$

$$volume : Endurant \times TimeRegion \mapsto Volume \qquad (2.8)$$

$$area : Endurant \times TimeRegion \mapsto Area \qquad (2.9)$$

Insofar as the quality regions are part of certain *structured space*, we assume the usual operators on them. For example, ratio scaled spaces like volumes and lengths are equipped with arithmetic operators for multiplication, addition and a fixed 0 element on atomic regions. Non-atomic regions are arbitrary subsets of the ratio scale. We furthermore assume that regions on the same level of measurement can be combined by operators into derived quality spaces [40].

Operations for identifying affordances in the environment. For example, operations identifying whether a place in a shop (endurant) affords to buy coffee (perdurant).

$$Affords[PerdurantType] : Endurant \times [PerdurantType] \mapsto Bool \qquad (2.10)$$

We conceive of an affordance as a boolean operation that decides whether a part of the environment (an endurant) allows for a simulated movement or action or resting. This means that such a simulation has successfully taken place in it, and that it is the minimal place necessary for the simulation. The endurant (e) thus identified gives rise to, and is part of (P), a larger medium (m) for the same kind of movement or action or resting (p):

$$Affords[PerdurantType](e,p) \rightarrow \exists m.Medium(m) \wedge P(e,m,time(p)) \qquad (2.11)$$

Note that if the involved perdurant is an action, e.g. *doMeasuring*, then the person acting also exists. Due to formula 2.3, an animate body is involved in generating this action. This body is logically necessary by surjectivity of the *do* operator, as required in the last paragraph of Sect. 2.3.3.

Operation for observing kinds of media in the environment. The *Medium* operator is parameterized with a type of endurant and perdurant. The idea is that the perdurant, e.g. a simulated action, is afforded by integral parts of the medium, while the endurant, e.g. an object, participates in that perdurant.

$$[PerdurantType][EndurantType]Medium : PhysicalObject \mapsto Bool \qquad (2.12)$$

A medium for motion, for example, has parts that afford motion of some type of endurant [see (2.13)]. Media for actions and restings have a corresponding usage. Note that media are rigid categories (not roles) just like bodies, because they cannot lose their affordance without disappearing. The underlying idea is that media are identified as integral wholes unified by affordances.[18] The following description captures only a necessary aspect of this idea, namely that a motion medium has a part that allows an endurant to perform a type of motion in it:

$$[MotionType][EndurantType]Medium(e) \rightarrow$$

$$\exists p, b, t.P(p,e,t) \wedge [EndurantType](b) \wedge$$

$$Affords[MotionType]\,(p,move[MotionType](b,t)) \qquad (2.13)$$

For example, a fluid is a medium with respect to a diving body,

$$Fluid(w) \leftrightarrow DivingBodyMedium(w) \qquad (2.14)$$

that is an integral part of the environment that affords a certain "low resistance" or "forceless" movement of this body.

As argued in [41], the notion of *place* can also be understood in terms of a medium, namely one which affords *containment* for animate bodies. Containment has many metaphorical meanings, but seems to be a central Gestalt schema of human cognition [42]. We conceive of it here as the human act of staying in a perceivable relation to a "container" in some physical sense. This can be a physical enclosure like a building or a conventionally demarcated place such as a bus station. For example, humans are inside a building if they stay in a certain relation to its inner surfaces, and they are at the bus station if they stay in a certain distance to the station sign.

$$Place(b) \leftrightarrow ContainingAnimateMedium(b) \qquad (2.15)$$

Operations for identifying features. Features [8] are perceivable parts of a body or medium identified with respect to a host object. An example is the opening of a funnel or the edge or surface of a table. In the first case, the feature, the opening, is not part of its host, the funnel, but part of the medium surrounding the funnel. But media can also be hosts for features. For example, a water body is a medium with a visible surface. Since there are different feature types an observer can distinguish, *identify* is an operator schema with a wildcard for subtypes of *Feature*. The most important feature is a *visible surface*, denoted by the type *Surface*:

$$identify[FeatureType] : PhysicalObject \mapsto [FeatureType] \qquad (2.16)$$

Due to surjectivity (Sect. 2.3.3), a feature always has a host body that generates it.[19] Features may be parts of bodies or media. We call the features that are part of a medium *OpenFeature*.

---

[18]More specifically, we conceive of them as maximal wholes self-connected by affordance relations among its parts [13].

[19]In DOLCE, while the authors seem to assume that features are existentially dependent on a host, it is left unspecified [8].

Operations for observing surface qualities. Many substances in the environment are specified by the surface quality of an object which is "made" of this substance. This may include texture and color, but also transparency. As an example, we introduce surface qualities as predicates over surfaces that allow to distinguish substances like snow from water:

$$Water : Surface \mapsto Bool \qquad (2.17)$$

$$Snow : Surface \mapsto Bool \qquad (2.18)$$

Operations for observing amounts. Amounts, like the amount of water contained in a bottle, must be observed based on other physical endurants with a unity criterion, e.g. physical objects, features or pluralities [37]. We have to identify an amount of water via the water unit that contains it at a certain moment in time, not vice versa. Furthermore, we track this amount through its various states, e.g. through merging or splitting into other objects. For example, when a statue is smashed, we track the amount of clay of this statue in terms of a remaining heap of clay. Sometimes, even the amount contained in a stable object keeps being exchanged, as in the case of a lake. In any case, amounts are first identified (and located) in terms of a temporal slice of an endurant (e.g., the statue before smashing, or a river part at t_n), and then need to be tracked through other temporal slices of other endurants (the lump after smashing, or another river part downstream at t_{n+1}). For the first task, we introduce a perceptual operation called *containsA*, which identifies the amount of matter of a (non-amount) endurant in a time moment. Because different endurants can contain the same amount of matter, and every amount is contained by an endurant, this operation is not injective, but surjective.[20]

$$containsA : NonAmountE \times TimeRegion \mapsto Amount \qquad (2.19)$$

$$P_A : Amount \times Amount \mapsto Bool \qquad (2.20)$$

$$Amount_{Fluid}(x) \leftrightarrow \exists y : Fluid, t.P_A(x, containsA(y,t)) \qquad (2.21)$$

The second operation for tracking amounts through endurant time slices is expressed by the parthood predicate P_A.[21] In contrast to parthood among non-amounts (2.1), parthood among amounts is extensional (for extensional mereologies, see [39]), so amounts are identical if they have the same parts, as argued in [37]. This means that subamounts cannot be exchanged. Since parts of an amount are not exchangeable,

---

[20]The operation corresponds to DOLCE's triadic "constitution" relation K between physical endurants restricted to amounts [8], but as constitution encompasses also abstract relations whose perceptual grounding remains unclear, we chose to use our own notion.

[21]Perceiving parthood among amounts could be based on following histories of endurant time slices (as in the case of rain drops falling into a lake), or on detecting exchange of matter among them (as in the case of flowing water). In every case, it is based on a very complex perceptual inference task on the part of the observer, which is not further discussed here.

P_A should be conceived as an atemporal relation. As in DOLCE, amounts and their time-slice containers are always *co-located*, and temporal parthood among endurants implies parthood of their contained amounts [8]. The two operators can be used to introduce *amount subtypes* such as the amount of a fluid $Amount_{Fluid}$ in (2.21) (we will refer to other amount subtypes in an equivalent way).

After having introduced and discussed unary perceptual types, their subsumption hierarchy and their interrelatedness via perceptual operations, we can now proceed to describe our first scenario in terms of such types and operations.

2.4 Technical Sensors

The challenge addressed in this section is how to describe measurement results of sensors in such a way that the observation process can be understood and in principle repeated by a user. It is particularly important that such a description is independent of technical implementations to be used for comparison, but specific enough to distinguish between different kinds of sensors. This has been identified as the major challenge by OGC's observation and measurement specification (O&M) [10], which states the need for an ontology describing properties:

> A schema for semantic definitions of property-types is beyond the scope of this specifi-
> cation. Ultimately this rests on shared concepts described in natural language. However,
> the value of the observed property is a key classifier for the information reported in an
> observation. Thus, in order to support such classification, for use in discovery and requests,
> an ontology of observable property-types must be available. [10]

2.4.1 Grounding Technical Sensors: Volumetric Flux and Volume Flow Rate

Many important measurements in hydrology, climatology and other geosciences are based on the idea of a flow of some fluid. Examples include measurements of precipitation conducted by a rain gauge and the flow rate of a river. Volumetric flux and volume flow rate are closely related properties, as each one can be derived from the other. In terms of SI units, volume flow rate is represented as the rate of volume flow across a given area, whereas volumetric flux is additionally normalized by this area:

$$vflowrate = \frac{m^3}{s} \qquad\qquad vflux = \frac{m^3}{m^2 * s} = \frac{m}{s} \qquad (2.22)$$

Both qualities are derived from the same kinds of measurement of volumes, areas and times. Volumetric flux sometimes can be reduced to the measurement of a length and a time (e.g. $\frac{mm}{h}$). From a semantic viewpoint, the problem is that this kind of description hides essential features, for example the fact that

Fig. 2.2 Identifying an open
feature (e.g. an opening) of a
funnel to get a cross-section
(the feature's location) and
its area

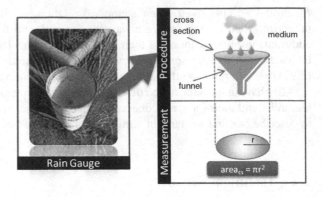

volumetric flux involves observing a certain kind of process in which a movement
of fluids is involved, and the distinction of this quality from arbitrary measurements
of velocities. Furthermore, SI units do not say much about other measurement
parameters involved: The location of measurement, or the object this quality inheres
in. In the following analysis, we focus on volumetric flux, but the same constituents
are involved for measuring volume flow rate.

2.4.2 Describing the Observation Procedures Underlying Volumetric Flux

Volumetric flux is obtained via the following procedures expressed in terms of
perceptual operations.

Identify a resting open feature. For measuring volumetric flux we need to identify
an amount of fluid moving through a cross-section. This cross-section is a location
in a medium for fluid amounts, and it may be identified by placing a collector
instrument, e.g. a funnel, which has an open feature indicating a free space, into such
a medium (see Fig. 2.2). It is the open feature that indicates the reference location
in the environment, and it is this location that the volumetric flux quality inheres
in. This feature needs to be an *OpenFeature*, as the cross-section must be part of
a medium, and not part of a body. The medium is one where *amounts of fluids*, for
example amounts of rain or flowing water, can move through. This can either mean
that the amount moves with its associated container,[22] as in the case of rain drops,
or it moves autonomously, as in the case of waterflow in a river.

– Input: Collector object (*collector*) and time during which the collector rests in
  the medium (*restingtime*).
– Output: A cross-section as a location in a medium for fluid amounts.

---

[22]These exist because of surjectivity of *containsA*.

Fig. 2.3 Identifying a
measuring action and
measuring its duration

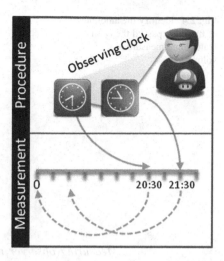

Identify an open feature...

$$identifyOpenFeature(collector) = openfeature \qquad (2.23)$$

...which rests for a certain time...

$$rest(collector, restingtime) = resting \qquad (2.24)$$

...in a medium which affords movement of fluid amounts.

$$\exists m.P(openfeature, m, restingtime) \wedge MotionAmount_{Fluid}Medium(m) \qquad (2.25)$$

The location of the collector's open feature is:

$$location(openfeature, restingtime) = crosssection \qquad (2.26)$$

Measure the area of the open feature. There are different operations for measuring areas of open features. For example, we can measure the radius of a maximal idealized circle located in the cross-section, inferring its area using π.

$$area(openfeature, restingtime) = area \qquad (2.27)$$

Identifying a measuring event and its duration.

- Input: The *observer*, the time during which the collector rests in the medium (*restingtime*), and the time interval of measuring (*measuretime*).
- Output: The duration of the measuring action performed during resting (see Fig. 2.3).

Fig. 2.4 Identify the amount of fluid that passed the open feature and measure its volume

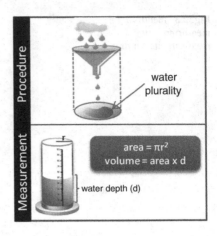

$$doMeasuring(observer, measuretime) = measuring \wedge$$

$$measuretime \subseteq restingtime \wedge$$

$$duration(measuretime) = duration \qquad (2.28)$$

Identify the amount of fluid that has passed the open feature during measuring. This can only be done by identifying a part of a fluid *fp* at a time *t*, which contains all and only those amounts that passed the open feature during measuring time.[23] Note that this does not necessarily mean that the human observer perceives the movement of an object, e.g. a unit of water. During measuring, the amount of fluid may be contained in a plurality of moving water units, e.g. in rain drops, but it may also be flowing inside one contiguous resting object, e.g. a river.

– Input: *measuretime, openfeature*, identification time *t*
– Output: *fp* (part of a fluid) identified at time *t* which contains all and only that amount of fluid that passed the open feature during measuretime.

Equation 2.29 describes what it means for a fluid part *fp* to contain *all* the fluid that passed the open feature (Fig. 2.4) during measuring (i.e. to be a "passing fluid container"): namely that *fp* needs to contain the amounts of all fluid parts *fp'* that where part of the open feature at some time *t'* during measuring (see Fig. 2.5):

$$\forall fp, t. PassingFluidCont(fp, t, measuretime, openfeature) \leftrightarrow$$

$$\exists w : Fluid.P(fp, w, t) \wedge$$

$$\forall fp', t'.(t' \subseteq measuretime \wedge (\exists z : Fluid.P(fp', z, t')) \wedge$$

$$P(fp', openfeature, t') \rightarrow P_A(containsA(fp', t'), containsA(fp, t))) \qquad (2.29)$$

---

[23]This is a slight oversimplification because it does not account for variations due to water loss or contamination.

Fig. 2.5 Explanation of (2.29)

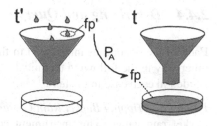

The sought operation must produce the minimal container in that sense, because it needs to contain *only* fluid of that kind. This restriction excludes unwanted amounts of fluid that did not pass the open feature:

$$\forall fp, t.passingFluidCont_{min}(t, measuretime, openfeature) = fp \leftrightarrow$$

$$\forall fp'.PassingFluidCont(fp', t, measuretime, openfeature) \rightarrow P(fp, fp', t) \qquad (2.30)$$

Measure the volume of this amount of fluid. Depending on the kind of fluid and the way the fluid part is identified, the procedure of measuring volume is different. We abstract from the specific implementation by a single perceptual operation of the volume quality.

$$volume(passingFluidCont_{min}(t, measuretime, openfeature), t) = volume \qquad (2.31)$$

2.4.3 Volumetric Flux in a Nutshell

A volumetric flux can be obtained by dividing the measured volume of the fluid amount by the cross-sectional area of the open feature and the measuring duration. Usual parameters of a particular volumetric flux value are its time of measuring, the location of the cross-section, and the time of recording *recT*. This is the time all required measurements are available so that the value can be recorded, which is in our case the time of identification of the amount of fluid:

$$volumetricFlux(measuretime, crosssection, recT)$$

$$= \frac{volume(passingFluidCont_{min}(recT, measuretime, openfeature), recT)}{area(openfeature) * duration(measuretime)}$$

$$(2.32)$$

If the reader substitutes all constants in this equation with perceptual function applications where possible, it can be seen that the observer, the resting, measuring and recording times, and the collector are the only free variables in this formula, which corresponds to our intuitive understanding of what setting is needed for observing this quality. It also makes clear that from an ontological viewpoint, the common parameters added to such a quality, like location and time, are only some of the entities involved in the context of measurement.

2.4.4 Describing and Querying Volumetric Flux Sensors

Each of the operations introduced in the last section is implemented in one or the other way in the following technical sensors. We will indicate the additional types needed in each case in the text.

Example 1: Tipping Bucket Rain Gauge. An example for a *rain gauge* is a tipping bucket rain gauge. This instrument comprises a funnel (i.e., *Funnel(collector)*) which collects the rain drops. The open feature is the opening (type *Opening*) of the funnel which allows rain drops to enter it. This means it is placed in a special medium *m* for "falling water", type $FallingAmount_{Waterunit}(m')$,[24] where the new type *Waterunit* means:

$$Waterunit(w) \leftrightarrow Fluid(w) \wedge Water(identifySurface(w)) \qquad (2.33)$$

The rain drops travel down the funnel and reach one of two 'small buckets' balanced on a fulcrum. The water amount passing the funnel during measuring is identified as the amount of raindrops accumulated in the full bucket (*fp*). When the rain drops fill one of two buckets located inside the gauge, the bucket tips and drains. The second bucket is positioned under the funnel for the next reading. The tipping event (*recT*) actuates a sealed reed switch which is detected by a data logger or telemetry system. The data logger records each individual tip of a bucket attributed to a specific time instant. The measuring perdurant (*measuring*) is simply the event between two tips. Since the bucket is filled and has a known volume (for example, each tip of the bucket represents 0.2 mm of rainfall), volumetric flux of rain can be computed with a constant volume but a varying measuring time.

Example 2: Snow Fall Measurement. Snow fall is a quality that seems to be conceptually easy, but turns out to be at least on the same level of complexity as rain gauge measurement. It is correctly conceptualized as the volume of snow accumulated on a piece of the ground surface during measuring time. The fluid amount medium is of type $FallingAmount_{Snowunit}(m')$ (derived from a medium of falling snowflakes, type *FallingSnowunitMedium(m)*), where:

$$Snowunit(s) \leftrightarrow Fluid(s) \wedge Snow(identifySurface(s)) \qquad (2.34)$$

The open feature in this case is a *Surface* feature: it is that part of the visible surface of the snow falling medium (where snow falling ends), which is located directly above the marked board on the ground surface (see Fig. 2.6). Its host is the board, which is at the same time the collector of snow (*Board(collector)*)). At the

---

[24]This medium is implied by a co-located medium for moving water units, type *FallingWaterunitMedium(m)*, with $P(m,m') \wedge P(m',m)$, since fluid amounts are co-located (and therefore move) with their containers. In order to draw this inference, a stronger theory of movement would be needed.

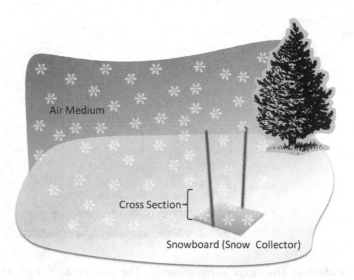

Fig. 2.6 Snow fall measurement. Starting of the measuring

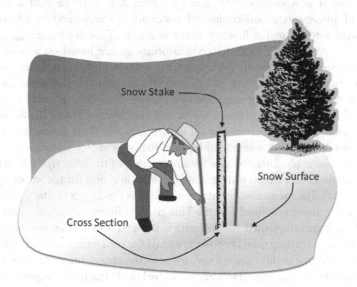

Fig. 2.7 Snow fall measurement. End of the measuring, recording

beginning of each measurement event, the board therefore has to be cleared from snow. The measuring event (*measuring*) is any major snow falling event. At the end of such an event (*recT*), the passed amount of snow is identified by that part of the snowpack (*fp*) which is right above the board after measuring time. Its volume is measured by sliding in a snow stake until it reaches the ground or snowboard and measuring depth (see Fig. 2.7). Since the area of the open feature happens to coincide with the bottom area of the accumulated snow pack, and thus disappears from the equation, volumetric flux becomes a velocity.

Fig. 2.8 River discharge
measurement

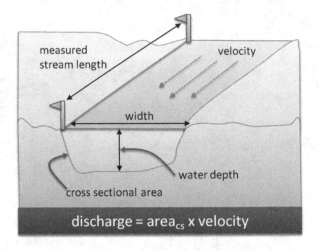

Example 3: Stream Discharge Measurement. The stream discharge or stream-flow is the volume of water passing a given cross-section along a stream in a unit time, so it is a volume flow rate (see Sect. 2.4.1). In case of a river, the medium of interest is a self-connected water unit, $Waterunit(m)$, which is co-located with a medium for flowing water amounts, $FlowingAmount_{Waterunit}(m')$, with $P(m,m') \wedge P(m',m)$.[25] Discharge measurements are based on a cross-section of this river. So the open feature (of type *Crosssect*) in this case is a cross-sectional part of the water unit of the river, and its host (the collector) is this water unit itself: $identifyCrosssect(m) = openfeature$, with $P(openfeature, m)$. In order to measure the area of this feature, the stream width and the average water depth measured at several locations in the cross-section are multiplied. Then a standard *traveling length* in the direction of flow has to be established. The amount of water passing the cross-section is identified by that volumetric part of the river (*fp*) which starts at the cross-section and extends orthogonally in flow direction for the traveling length (see Fig. 2.8).[26] The measuring event (*measuring*) is the time a floating object takes to travel along this predefined length. This time indirectly measures how long it takes to fill up the established volume with water and how fast an amount of water is moving. The recording time *recT* is the end of the measuring event, at which *fp* is filled. So the volume in this case is fixed, whereas the measuring duration is variable.

Descriptions like the ones above are on a level of abstraction adequate to express useful differences among sensors based on property or quality types. They could improve queries by abstracting from the technical level of O&M in describing

---

[25] In this case, the medium for flowing amounts of water is not implied by the other medium, since the water unit itself is not moving. This illustrates the use of distinguishing these different kinds of media.

[26] This is a simplification, since it is assumed that this volume is simply the orthogonal projection of the cross-section in travelling direction.

which kinds of qualities and objects are observed. For example, we can compare volumetric flux sensors on the level of types of fluid media, for example on the level of *flowing objects* (amounts vs. bodies of snow or units of water) or on the level of *movement types* (falling or flowing), or based on whether the open feature is fully "drained" in the fluid (river) or not (precipitation). Different implementations of rain gauges (standard rain gauge, tipping bucket, a.s.o) can be subsumed under a common type.

Furthermore, the instruments and actions (i.e., the resources) involved in such a measurement can be compared in detail. We see that in some cases, a collector instrument (a body with a funnel) needs to be placed somewhere, whereas in other cases the collector is just an existing water body. The operation for identifying amounts and measuring volumes and areas is implemented differently in each case. The measuring event requires to be aligned with different other observations by the observer, like measuring the traveling time of water at a fixed distance, or the duration of a snowfall event. In other cases the event is triggered automatically, as in case of the tipping bucket. The temporal intervals of these measuring events are huge or small and have fixed or open duration depending on these implementations.

How can such queries be implemented? In our scenario we mostly needed sentences with *ground terms* (terms without variables) or existentially quantified variables. This makes it easy to substitute all existentially quantified variables and all function applications by constants, and all functions by relations, in order to represent the resulting ground sentences in a relational scheme, e.g. a relational database or RDF. We then also have to replace Definition 2.30, which employs a universal quantifier, by a primitive relation, taking a loss on expressiveness. Examples how such a relational scheme can be translated into RDF and how it can be queried with appropriate languages is shown in Sect. 2.5.3.

2.5 Volunteered Geographic Information

In the previous section, we have discussed how technical sensors can be semantically referenced based on the underlying measurement and observation procedures. While this approach targets semantic annotation of sensors and observations at a technical level, the amount of geographic information that is produced by non-technical sensors has grown rapidly in recent years. *Citizens as sensors* [7] produce a range of different kinds of geographic information, with the community-generated world map in the OSM project being the most prominent example. Other examples of such VGI include maps for local natural hazards (such as bush fires) where the authorities have difficulty in providing up-to-date maps, or services on the Social Web where the location information is produced as a by-product, such as geo-tagged pictures, blog posts or tweets.[27]

---

[27] See http://twitter.com.

On the surface, the observation methods applied by human observers that produce such geographic information seem to be fundamentally different from those underlying the technical sensors discussed in Sect. 2.4. In this section, we argue that the volunteered observation process can best be described based on the human perception and simulation of possible interactions with the environment. In this respect, technical and human observations have a common root: They can both be described in terms of perceptual types, e.g. affordances, in a shared environment. We show how affordances can be employed to improve the tagging of geographic features. OSM serves as a working example in this section, as it provides the largest collection of user-generated geographic information. We discuss an approach to semantic referencing of POI in OSM.

2.5.1 OpenStreetMap: Describing the Semantics of POIs

In a German OSM mailing list,[28] people recently discussed how to tag a POI where you can mail as well as pick up letters and parcels. Since a lot of automated boxes offering this functionality have been installed in Germany[29] over the past years, people recognized that the commonly used label post_box does not specify what is really of interest for the user: can he or she mail or pick up letters and parcel?

POI in OSM are currently annotated with key-value pairs such as amenity= post_office. This combination of keys and values (also referred to as tags) is not structured in any way. Users can freely assign any tags they consider useful for a POI when they add or edit it. The idea is to reach a consensus on the tags to use by discussions on the OSM wiki and different mailing lists.[30] The tags assigned this way, however, can make it difficult for users to properly annotate their POIs. Places often do not fit into just one of the category values defined by the OSM keys.

For example, many cafés in Europe are open late and also serve alcohol, so that they would be better described as bars in the evening. Annotating such POIs with either amenity=cafe or amenity=bar therefore only tells half the truth and hides useful information from OSM users. Such conflicts can often be observed, for example when bookstores include cafés, when kiosks serve as pick-up points for parcel companies, or when fully annotating these POIs with all relevant information is only possible with workarounds. Post offices offer a number of different services in Germany, including banking facilities (provided by the Postbank), ATM machines, the opportunity to buy stationary in addition to the

---

[28] See http://lists.openstreetmap.org/pipermail/talk-de/2008-February/007487.html.

[29] See http://www.dhl.de/en/paket/privatkunden/packstation.html.

[30] In practice, however, the decision on the tags to use is largely influenced by the developers of the different map renderers and editors. They decide which tags are picked for display on the map, and which symbols are chosen for them. Since seeing "your" POI on OSM is a major incentive to contribute, users obviously pick certain tags to describe their POIs.

common services available at a post office. These come in a number of different combinations, depending on the size and location of the office. Describing such a POI only as `amenity=post_office` does not give credit to all these different functionalities and makes it hard for other users to figure out what kind of services they can expect at this place.

One problem of the OSM data scheme lies in the fact that a place can have a variety of affordances, which may be of varying importance depending on the user. POI in OSM in contrast allow for only one value per `amenity` key, wich implies an a-priori choice. Another problem is that the categories suggested as values are too abstract and ambiguous to give a clear idea of how they should be observed.

To overcome these problems of the current OSM tagging approach, we propose to use tags grounded in affordances. Similar to the affordance-based specification of observation procedures introduced in the previous section, this approach is based on reproducible observation of different functionalities offered by the POIs. We use several examples to illustrate how affordance-based tagging can help with a solution.

2.5.2 Describing the Observation Procedures Underlying POI Affordances

We suggest to conceive of POIs as places equipped with a (potentially long) list of action affordances offered by its parts (compare [43]). These affordances account for the "interest" in a POI. Note that we do not suggest that there is any intrinsic priority ordering among them, for example some prior use. Our view also follows the open world assumption that what volunteers do not know, i.e., the affordances they may not have observed, may nevertheless be existent. In the following, we go through several POI examples (assuming that the unbound variable *poi* denotes any particular one) to demonstrate how the observations underlying POIs could be described.

Restaurants, cafés and bars are places, and some of their parts, namely tables and chairs, afford to eat, drink and talk to each other. There may be also other parts such as a bar, toilets and entrances. Definitions of these place categories would be very complex because of their graded prototypical nature. It is also not recommendable to suggest some "major" type of usage, because restaurants can be used as cafés and vice versa, depending on the intentions of the observer. In some situation, a restaurant may be primarily a place to find a restroom or an open wifi network. It seems therefore more reasonable to add functional descriptions of a *poi* place using affordance operators differentiated by action types, while keeping involved persons, things and times implicit (existentially quantified). Sometimes, explicit parameters may be useful. For example, t in the following example indicates the time of observation of the affordance.

Restaurant, café or bar:

$Place(poi) \wedge \exists eatingplace.P(eatingplace, poi) \wedge$

$\exists somebody, something, t.Affords(eatingplace, doEat(somebody, something, t)) \wedge$

$\exists drinkingplace.P(drinkingplace, poi) \wedge$

$\exists somebody, t.Affords(drinkingplace, doDrinkAlcohol(somebody, t))$ \hfill (2.35)

Supermarkets. We can distinguish supermarkets from restaurants by asserting that they allow to buy food, but not to eat it there. Since categories like food are themselves observed via affordances such as eatability, this nicely fits into our type schema:

Supermarket:

$Place(poi) \wedge \exists pointofsale.P(pointofsale, poi) \wedge$

$\exists somebody, something, t.Affords(pointofsale, doBuy(somebody, something, t)) \wedge$

$\exists somewhere, t.Affords(somewhere, doEat(somebody, something, t))$ \hfill (2.36)

Parking lots are places that allow to place vehicles, which are things to drive with:

Parking lot:

$Place(poi) \wedge \exists lot.P(lot, poi) \wedge \exists car, t.Affords(lot, rest(car, t)) \wedge$

$\exists somewhere, somebody, t.Affords(somewhere, doDrive(somebody, car, t))$ \hfill (2.37)

The descriptions above not only highlight which contextual entities need to be observed, but also which entities can be inferred if a POI is being observed.

Evidently, this approach is in stark contrast to the simple key-values pairs currently used in OSM. An affordance-based tagging potentially leads to more complete and appropriate descriptions of POI and hence facilitates retrieval, as outlined in the following section. The complexity of the underlying formalization, however, may hamper users from contributing to the map. This approach hence requires user interfaces that hide the complexity from OSM users. Furthermore, the complexity of an observed affordance can be hidden by definitions. The general schema of a POI affordance could be shortened, e.g., by the defined binary predicate *HasObservedAction*:

$$\forall poi, action.(poi)HasObservedAction(action) \leftrightarrow$$

$$Place(poi) \wedge \exists p.P(p, poi) \wedge Affords(p, action) \hspace{2em} (2.38)$$

2.5.3 Querying and Visualizing Affordance-Based POI Tags

The immediate advantage of affordance-based descriptions becomes evident when querying over the potential functions and their links to existing entities. In this case, querying for a specific function would result in all unexpected and expected uses of a POI. So a query for the function of drawing money would return bank offices as well as ATM machines. Similarly to the use case in Sect. 2.4.4, the sentences above contain only ground terms or existentially quantified variables, and so they can be easily represented and queried in any relational database: Simply replace the existential quantifiers by constant names, convert all terms (constant names and function applications) into keys of a table (one table for every type), and then convert every atomic formula to a table row (one table for every predicate) referencing constant names with foreign keys. It is similarly possible to convert this scheme into a set of RDF triples linked with OWL concepts in order to directly annotate OSM data. In this case, SPARQL technology [44] as well as a RESTful interface [45] can be used to query afforded actions.

The following RDF snippet demonstrates the annotation process for a café in Münster. It uses a vocabulary including the RDF predicate hasObservedAction (as defined above) and others from a hypothetical file http://foo.bar/poi. Note that the affordance-based tagging does not render existing key-value pairs useless, but should rather complement them.[31] Categories which are currently assigned manually could be defined based on sets of affordances, as outlined in Sect. 2.5.2. POIs falling into a specific category could be automatically tagged based on the assigned affordances. In order to keep the set of assigned affordances consistent, recommendations based on existing affordances can be shown to the users [46].

```
@prefix poi: <http://foo.bar/poi#>.
@prefix osm: <http://osm.org/browse/node/>.
@prefix rdf: <http://www.w3.org/1999/02/22-rdf-syntax-ns#>.
osm:740777363
        poi:hasName "Teilchen und Beschleuniger";
        poi:hasObservedAction
                <poi:doDrinkAlcohol_37>, <poi:doDrinkCoffee_21>,
                <poi:doEat_42>;
        poi:isAmenity "bar","cafe".
        ...
```

The observed actions, e.g. doDrinkCoffee, are instances of more generic action types, e.g. DrinkCoffee, as specified in the following excerpt. These action types are rdfs:subClassOf the generic type Action. The specific instances are observed at specific time instances, which would allow to infer temporally restricted

---

[31]For an implementation of affordance-based tagging, the OSM restriction of having only one value per key would have to be loosened (which is not an issue in RDF).

action potentials (e.g., to deduce opening hours or when food is served only up to a certain time).[32]

```
@prefix poi: <http://foo.bar/poi#>.
@prefix rdf: <http://www.w3.org/1999/02/22-rdf-syntax-ns#>.
<poi:doDrinkCoffee_21>
    poi:observedAt "2010-06-05T18:00:00-5:00"^^<xsd:date>;
    rdf:type <poi:DrinkCoffee>.
```

The following SPARQL code shows a sample query for OSM POIs that afford to eat (i.e., to serve food) within the next two hours. We assume a conversion of the date time at which this action was observed to "hours of day":

```
PREFIX poi: <http://foo.bar/poi#>
PREFIX time: <http://www.w3.org/2006/time#>
SELECT ?cafe ?eat ?now ?inTwoHours
WHERE {
  ?y poi:isAmenity ?cafe;
  ?y poi:hasObservedAction ?eat;
  ?eat poi:observedAtHour ?t;
  ?t time:after ?now;
  ?t time:before ?inTwoHours.
}
```

This approach would also enable a different way to hook OSM data into the Linked Data Cloud, as proposed by the Linked GeoData initiative [47], and semantically enable OSM data for the exchange of spatial information [48].

Evidently, affordance lists are difficult to visualize by icons when rendering the map. Since the affordance-based approach does not exclude existing tags in OSM, the renderer's categories based on these tags can still be represented by the existing icons,[33] so that no changes are required at this point. A small extension to the mapping interface would make the novel query functionality accessible for the user: if one searches for functionalities on the map (e.g. "draw money"), the POIs offering this functionality could simply be highlighted. Hence, there is no need for special icons for every affordance.

2.6   Conclusions and Future Work

In this chapter, we have made a proposal for describing the semantics of geodata on an experiential level, as a means to resolve its inherent abstraction and grounding problems. These problems become manifest in that geosensors are inadequately described on the level of SI units as well as on the level of instruments, and that useful and consistent tagging of VGI is a matter of choosing a level of categories with clear interpretations.

---

[32] See http://www.w3.org/TR/owl-time/ and its RDF resource under http://www.w3.org/2006/time.
[33] See http://wiki.openstreetmap.org/wiki/Rendering.

We propose to add semantic references to such data as a way of enabling users to reproduce the underlying observation processes. To this end, we suggested an operational view of human perception, including basic perceptual types such as media, bodies, features, motions, actions, amounts and quality regions, which can be aligned with top-level categories of DOLCE, and which have criteria of individuation rooted in Gestalt perception capabilities. The proposed types are linked by perceptual operations, e.g. action-, motion-, affordance-, and feature-detectors, which account for the generative dependence among them, and which are also alignable with DOLCE. Criteria of individuation were not discussed here, but might be given in terms of integral wholes, as proposed in [3, 20].

Our main idea is that affordances and the equipment of the meaningful environment, understood as perceptual types, provide a firm basis for semantic referencing of geodata. We demonstrated that volumetric flux sensors, e.g. rain gauges, can be described with the same formal apparatus as POI in volunteered information. Thereby, the variety of entities involved in a measurement, as well as the commonalities between instrumental implementations show up. We furthermore found that POI can be adequately modelled as places with lists of affordances related via perceptual operations.

The formal framework for semantic referencing sketched and illustrated in this chapter allows to formulate typed first-order theories about geosensors and VGI, of which we have only scratched the surface. In order to do that, the proposal needs a deeper formal elaboration. The main questions are what sets of axioms consistent with the ones in DOLCE should be added, and what further media, motion, action and quality subtypes are necessary to describe a given domain.

The framework can be used in its current form as a guideline for annotating VGI and geosensors. As we have discussed, subsets of a typed reference theory can be translated into RDF, facilitating queries over VGI and geosensors. It is straightforward to convert the sentences used in the examples into ground terms, substituting existentially quantified entities with object constants, and complex functions with primitive relations. The resulting data scheme is just a simple RDF graph of instances. It can be efficiently handled by standard triple stores, indices, and query engines based on SPARQL.

We consider all these options, as well as the required implementations, as part of future work.[34]

Acknowledgments This research has been funded by the Semantic Reference Systems II project (DFG KU 1368/4-2), the SimCat project (DFG Ra1062/2-1 and Ja1709/2-2, see sim-dl.sourceforge.net), and the International Research Training Group on *Semantic Integration of Geospatial Information* (DFG GRK 1498, see irtg-sigi.uni-muenster.de).

---

[34] A similar framework was used to implement a tool for evaluating and querying road network junctions in OSM (compare [20]). The tool is freely available as JOSM plugin under http://wiki.openstreetmap.org/wiki/JOSM/Plugins/JunctionChecking.

References

1. Köhler, W.: Gestalt Psychology. An Introduction to new Concepts in Modern Psychology. Liveright, New York (1992)
2. Tomasello, M.: The cultural origins of human cognition. Harvard University Press, Cambridge, MA (1999)
3. S. Scheider, F.P., Janowicz, K.: Constructing bodies and their qualities from observations. In: Formal Ontology in Information Systems. Proc. of the 6th Int. Conf. IOS Press, Amsterdam (2010) 69–87
4. Scheider, S.: The case for grounding databases. In: 3rd Int. Conf. on GeoSpatial Semantics. Volume 5892 of LNCS. Springer, Berlin (2009) 44–62
5. Harnad, S.: The symbol grounding problem. Physica D: Nonlinear Phenomena **42**(1-3) (1990) 335–346
6. Smith, B.: The limits of correctness. ACM SIGCAS Computers and Society **14**(1) (1985) 18–26
7. Goodchild, M.: Citizens as sensors: the world of volunteered geography. GeoJournal **69**(4) (2007) 211–221
8. Masolo, C., Borgo, S., Gangemi, A., Guarino, N., Oltramari, A.: Wonderweb deliverable d18: Ontology library, Trento, Italy (2003)
9. Neuhaus, F., Grenon, P., Smith, B.: A formal theory of substances, qualities and universals. In Varzi, A., Vieu, L., eds.: Proc. of the 3rd Int. Conf. on Formal Ontology in Information Systems (FOIS-04). IOS Press (2004) 49–59
10. Cox, S., et al.: Observations and measurements. part 1 - observation schema (ogc 07-022r1) (2007)
11. Probst, F.: An ontological analysis of observations and measurements. In Raubal, M.e.a., ed.: Proc. of the 4th. International Conference on Geographic Information Science (GIScience 2004). Lecture notes in computer science. Springer, Berlin (2006) 304–320
12. Kuhn, W.: Semantic reference systems. Int. J. Geogr. Inf. Sc. **17** (5) (2003) 405–409
13. Scheider, S., Janowicz, K., Kuhn, W.: Grounding geographic categories in the meaningful environment. In K.S. Hornsby, C. Claramunt, M.D., Ligozat, G., eds.: Spatial Information Theory, 9th Int. Conf., COSIT 2009, Proc. Springer, Berlin (2009) 69–87
14. Bishr, Y.: Overcoming the semantic and other barriers to gis interoperability. International Journal of Geographical Information Science **12** (1998) 299–314
15. Kuhn, W.: Modeling vs. Markup. Semantic Web Journal (under review)
16. Gibson, J.: The ecological approach to visual perception. Houghton Mifflin, Boston (1979)
17. Shaw, R., Turvey, M., Mace, W.: Ecological psychology: The consequence of a commitment to realism. In Weimer, W., Palermo, D., eds.: Cognition and the Symbolic Processes. Volume 2. Lawrence Erlbaum Associates, Inc., Hillsdale, NJ (1982) 159–226
18. von Glasersfeld, E.: Radical Constructivism: A Way of Knowing and Learning. The Falmer Press, London (1995)
19. Pylyshyn, Z.: Things and Places. How the Mind Connects with the World. The MIT Press, Cambridge, Massachusetts (2007)
20. Scheider, S., Kuhn, W.: Affordance-based categorization of road network data using a grounded theory of channel networks. Int. J. Geogr. Inf. Sc. **24**(8) (2010) 1249–1267
21. Guarino, N., Welty, C.: Identity, unity and individuality: Towards a formal toolkit for ontological analysis. In Horn, W., ed.: ECAI 2000 Proc. Volume 54 of Frontiers in artficial intelligence and applications. IOS-Press, Amsterdam (2000) 219–223
22. Sanders, J.: An ontology of affordances. Ecological Psychology **9** (**1**) (1997) 97–112
23. Scarantino, A.: Affordances explained. Philosophy of Science **70** (2003) 949–961
24. Turvey, M.: Affordances and prospective control: An outline of the ontology. Ecological Psychology **4** (**3**) (1992) 173–187
25. Warren, W.: Perceiving affordances: Visual guidance of stair climbing. Journal of Experimental Psychology: Human Perception and Performance **10**(5) (1984) 683–703

26. Stoffregen, T.A.: Affordances as properties of the animal-environment system. Ecological Psychology **15 (2)** (2003) 115–134
27. Oudejans, R., Michaels, C., van Dort, B., Frissen, E.: To cross or not to cross: The effect of locomotion on street-crossing behavior. Ecological Psychology **8**(3) (1996) 259–267
28. Barsalou, L.C.: Perceptual symbol systems. Behavioral and Brain Sciences **22** (1999) 577–660
29. Ortmann, J., Kuhn, W.: Affordances as qualities. In: Formal Ontology in Information Systems. Proc. of the 6th Int. Conf. LNCS. IOS Press, Amsterdam (2010) 117–130
30. Turner, P.: Affordance as context. Interacting with Computers **17**(6) (2005) 787–800
31. Brodaric, B., Probst, F.: DOLCE ROCKS: Integrating Foundational and Geoscience Ontologies. In: AAAI 2008 Spring Symposia, Stanford University, California (2008)
32. Klien, E.: Semantic Annotation of Geographic Information. PhD Thesis, Institute for Geoinformatics, University of Münster, Germany (2008)
33. Gangemi, A., Guarino, N., Masolo, C., Oltramari, A., Schneider, L.: Sweetening ontologies with DOLCE. In: EKAW '02: Proceedings of the 13th International Conference on Knowledge Engineering and Knowledge Management. Ontologies and the Semantic Web, London, UK, Springer-Verlag (2002) 166–181
34. Probst, F.: Observations, measurements and semantic reference spaces. Applied Ontology **3**(1) (2008) 63–89
35. Kuhn, W.: A functional ontology of observation and measurement. In Janowicz, K., Raubal, M., Levashkin, S., eds.: Third Workshop on Geosemantics (GeoS 2009), Mexico City, 3-4 December 2009. Volume 5892 of Lecture Notes in Computer Science., Berlin Heidelberg, Springer (2009) 26–43
36. Guarino, N., Welty, C.: A formal ontology of properties. In Dieng, R., Corby, O., eds.: Knowledge Acquisition, Modeling and Management: 12th International Conference, EKAW 2000. Volume 1937 of Lecture notes in computer science. Springer, Berlin (2000) 97–112
37. Guizzardi, G.: On the representation of quantities and their parts in conceptual modeling. In: Formal Ontology in Information Systems. Proc. of the 6th Int. Conf. LNCS. IOS Press, Amsterdam (2010) 103–116
38. Rosch, E.: Principles of categorization. In Rosch, E., Lloyd, B., eds.: Cognition and categorization. Lawrence Erlbaum (1978) 27–48
39. Casati, R., Varzi, A.: Parts and places: The structures of spatial representation. MIT Press, Cambridge, Mass. (1999)
40. Probst, F.: Observations, measurements and semantic reference spaces. Journal of Applied Ontology **3(1-2)** (2003) 63–89
41. Scheider, S., Janowicz, K.: Places as media of containment. In: Proceedings of the 6th International Conference on Geographic Information Science (extended abstract, forthcoming). (2010)
42. Johnson, M.: The body in the mind: The bodily basis of meaning, imagination, and reason. University of Chicago Press, Chicago (1987)
43. Jordan, T., Raubal, M., Gartrell, B., Egenhofer, M.: An affordance based model of place in gis. In: Proceedings of 8th International Symposium on Spatial Data Handling (SDH'98), International Geographic Union (1998) 98–109
44. Prud'hommeaux, E., Seaborne, A.: SPARQL Query Language for RDF. W3C Recommendation 15 January 2008, available from http://www.w3.org/TR/2008/REC-rdf-sparql-query-20080115/
45. Fielding, R.: Architectural Styles and the Design of Network-based Software Architectures. PhD thesis, University of California, Irvine, USA (2000)
46. Trame, J.: Recommending POI tagging in Openstreetmap by using the co-occurrence of tags. Bachelor's thesis, Institute for Geoinformatics, University of Mnster (2010)
47. Auer, S., Lehmann, J., Hellmann, S.: LinkedGeoData – adding a spatial dimension to the web of data. In: Proc. of 7th International Semantic Web Conference (ISWC). (2009)
48. Janowicz, K., Schade, S., Bröring, A., Keßler, C., Maué, P., Stasch, C.: Semantic enablement for spatial data infrastructures. Transactions in GIS **14**(2) (2010) 111–129

Chapter 3
SPARQL-ST: Extending SPARQL to Support Spatiotemporal Queries

Matthew Perry, Prateek Jain, and Amit P. Sheth

Abstract Spatial and temporal data is plentiful on the Web, and Semantic Web technologies have the potential to make this data more accessible and more useful. Semantic Web researchers have consequently made progress towards better handling of spatial and temporal data.SPARQL, the W3C-recommended query language for RDF, does not adequately support complex spatial and temporal queries. In this work, we present the SPARQL-ST query language. SPARQL-ST is an extension of SPARQL for complex spatiotemporal queries. We present a formal syntax and semantics for SPARQL-ST. In addition, we describe a prototype implementation of SPARQL-ST and demonstrate the scalability of this implementation with a performance study using large real-world and synthetic RDF datasets.

3.1 Introduction

Nearly all human activity is rooted in space and time, and increasing amounts of spatial and temporal data are appearing on the Web. Examples include spatial and temporal data about tracking hurricanes and aquatic animals.[1,2] We have also seen increasing amounts of user-generated geospatial metadata created with geotagging vocabularies such as GeoRSS. The number of Web mashups created with public map services alone is a testament to the usefulness of maps and spatial data in a variety of applications. These real-world scenarios motivate us to argue that current tools for managing Semantic Web data must be extended to better handle spatial and temporal data.

---

[1] http://weather.unisys.com/hurricane/index.html.

[2] http://whale.wheelock.edu/whalenet-stuff/stop_#cover.html.

M. Perry (✉)
Oracle, 1 Oracle Drive, Nashua, NH 03062, USA
e-mail: matthew.perry@oracle.com

N. Ashish and A.P. Sheth (eds.), *Geospatial Semantics and the Semantic Web: Foundations, Algorithms, and Applications*, Semantic Web and Beyond 12, DOI 10.1007/978-1-4419-9446-2_3, © Springer Science+Business Media, LLC 2011

Researchers have made initial progress in this direction. Gutierrez et al. proposed Temporal RDF Graphs to model temporal aspects of RDF triples [8, 9]. The RDF statement is extended in this model from a triple to a quad where the fourth element is the valid time of the RDF statement. There has also been significant research inspired by the Geospatial Semantic Web vision [6]. An architecture of ontologies for the Geospatial Semantic Web has been proposed [14], and a variety of tools and systems to manage spatial data on the Semantic Web have been introduced [15, 22, 27]. In addition, groups such as the W3C Geospatial Incubator Group [17] have pursued standard ontologies for geospatial data.

Query language support for spatial and temporal RDF data is currently lacking. SPARQL [23] has recently emerged as the W3C-recommended query language for RDF data, but, to date, no extensions of SPARQL to support complex spatial and temporal queries exist. This chapter proposes SPARQL-ST, an extension of SPARQL that supports queries over spatiotemporal RDF graphs (i.e. temporal RDF Graphs that contain spatial objects). Consider the SPARQL-ST query below. This query selects all politicians (and their tenure) that represent a congressional district that is inside a given polygon.

```
SELECT ?p, %g, intersect(#t1, #t2, #t3, #t4)
    WHERE {
    ?p usgov:hasRole ?r #t1 .
    ?r usgov:forOffice ?o #t2 .
    ?o usgov:represents ?c #t3 .
    ?c stt:located_at %g #t4 .
    SPATIAL FILTER (inside(%g, GEOM(POLYGON ((-75.14 40.88,
    -70.77 40.88, -70.77 42.35, -75.14 42.35,-75.14 42.35,
    -75.14 40.88)))))
```

In addition to normal SPARQL variables (denoted with a ? prefix), SPARQL-ST introduces a spatial variable type (denoted with a % prefix) and a temporal variable type (denoted with a # prefix). Spatial variables represent complex spatial features rather than a single URI, and the concept of a mapping is extended so that spatial variables map to a set of triples that represent a spatial feature. The spatial variable %g is used in the query above to represent the spatial extent of a congressional district. Temporal variables map to time intervals rather than a URI and can appear in the quad position of what we term a spatiotemporal triple pattern. Temporal variables are used in the example query to retrieve the valid time of each temporal RDF statement. In addition, SPARQL-ST allows computation of derived time intervals. For example, the query above computes the interval intersection of four time intervals to derive the valid time of the entire triple pattern. SPARQL-ST also introduces *SPATIALFILTER* and *TEMPORALFILTER* expressions to filter results using spatial and temporal conditions. The query above applies a filtering condition to the spatial extent of each congressional district.

With the objective of providing better support for spatiotemporal queries over Semantic Web data, this work makes the following contributions:

1. A formal syntax and semantics for the SPARQL-ST query language.
2. A prototype implementation of SPARQL-ST built on top of a relational database management system.

3. A performance evaluation of the prototype system using both synthetic and real-world RDF datasets.

The remainder of the chapter is organized as follows. Section 3.2 presents the RDF data model and approaches for modeling spatial and temporal data in RDF. Section 3.3 introduces the SPARQL-ST query language by defining its formal syntax and semantics. Section 3.4 describes our prototype implementation of SPARQL-ST, and Sect. 3.5 evaluates the scalability of our prototype using synthetic and real-world RDF datasets. Related work is discussed in Sect. 3.6. Finally, Sect. 3.7 gives conclusions and discusses directions for future work.

3.2 Modeling Approach

We give details of our approach for modeling spatial and temporal data using RDF in this section. We incorporate temporal information using Temporal RDF Graphs [8,9], and we present an ontology based on the Open Geospatial Consortium (OGC) Geographic Modeling Language (GML) specification to model spatial features. Temporal RDF triples are encoded using standard RDF reification. Our formal definition of SPARQL-ST depends on this modeling approach, so we present the specifics of our modeling approach (first described in [21]) as a prerequisite to our SPARQL-ST definition in the next section. We first formally define the RDF model and Temporal RDF graphs and then present our ontology for spatial features. Although SPARQL-ST currently depends on a particular serialization of temporal RDF and spatial ontology, the concepts of SPARQL-ST are equally applicable to other temporal RDF serializations and other spatial ontologies.

3.2.1 RDF

RDF has been adopted by the W3C as a standard for representing metadata on the Web. The RDF data model is defined as follows. Let U, L and B be pairwise disjoint sets of URIs, literals and blank nodes, respectively. The union of these sets $U \cup B \cup L$ is referred to as the set of RDF Terms RT. An *RDF triple* is a 3-tuple $(s, p, o) \in (U \cup B) \times U \times RT$ where s is the *subject*, p is the *property* and o is the *object*. A set of RDF triples is referred to as an *RDF Graph*, as RDF can be represented as a directed, labeled graph where a directed edge labeled with the property name connects a vertex labeled with the subject name to a vertex labeled with the object name. RDF Schema (RDFS) [4] provides a standard vocabulary for describing the classes and relationships used in RDF graphs and consequently provides the capability to define ontologies.

A set of entailment rules are also defined for RDF and RDFS [10]. Conceptually, these rules specify that an additional triple can be added to an RDF graph if the graph contains triples of a specific pattern. Such rules describe, for example, the transitivity of the *rdfs:subClassOf* property.

3.2.2   Temporal RDF

In order to analyze the temporal properties of relationships in RDF graphs, we need a way to record the temporal properties of the statements in those graphs, and we must account for the effects of those temporal properties on RDFS inferencing rules. Gutierrez et al. [8, 9] introduced the notion of temporal RDF graphs for this purpose.

Temporal RDF graphs model linear, discrete, absolute time and are defined as follows [8]. Given a set of discrete, linearly ordered time points T, a *temporal triple* is an RDF triple with a temporal label $t \in T$. A statement's temporal label represents its valid time. The notation $(s,p,o) : [t]$ is used to denote a temporal triple. The expression $(s,p,o) : [t_1,t_2]$ is a notation for $\{(s,p,o) : [t] \mid t_1 \leq t \leq t_2\}$. *A temporal RDF graph* is a set of temporal triples.

Let us consider the politician Bill Clinton who was governor of Arkansas from 11 January 1983 until 12 December 1992 and president of the United States from 20 January 1993 until 20 January 2001. This would yield the following triples: (*Bill_Clinton, holds_office, AR_Governor*) : [01:11:1983, 12:12:1992], (*Bill_Clinton, holds_office, US_President*) : [01:20:1993, 01:20:2001].

We must also account for the effects of temporal labels on RDFS inferencing rules. To incorporate inferencing into temporal RDF graphs, a basic arithmetic of intervals is needed to derive the temporal label for inferred statements. For example, interval intersection would be needed for *rdfs:subClassOf* (e.g., (*x, rdfs:subClassOf, y*) : [1,4] $\wedge$ (*y, rdfs:subClassOf, z*) : [3,5] $\Rightarrow$ (*x, rdfs:subClassOf, z*) : [3,4]).

We use RDF reification to associate time intervals with RDF statements to realize temporal RDF graphs. RDF reification is a construct in RDF that allows one to make statements about statements, so we can assert that a given RDF statement has a given valid time. We use a portion of the OWL-Time ontology [12] to model the time intervals, and a new property *temporal* asserts that the reified statement is valid during the given time interval. Figure 3.1 illustrates this approach.

3.2.3   Spatial Ontology

Spatial features are complex types that need to be fully modeled with a spatial ontology. Fortunately, there is movement towards standard ontologies for spatial geometries, for example work done as part of the OGC Semantic Web Interoperability Experiment [18] and the W3C geo incubator group [17]. The existing OGC GML specification serves as an excellent basis for these ontologies as discussed in [1] and [14]. We propose a spatial ontology based on the GeoRSS GML specification [26]. The ontology models 2-dimensional spatial geometries and associated spatial reference system information. Figure 3.2 illustrates the RDFS representation of this ontology.

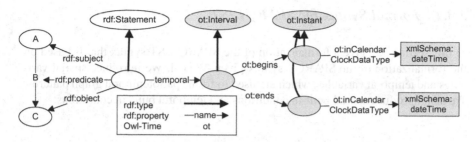

Fig. 3.1 Temporal reification of the RDF statement (*A B C*). Constructs from the Owl-Time ontology are shown in *gray*

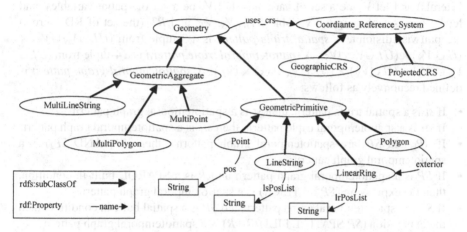

Fig. 3.2 GeoRSS GML-based ontology modeling basic spatial geometries. Note that Geometric Aggregates contain collections of their respective Geometric Primitives (e.g., multipolygon contains a collection of polygons). These relations and attributes of Coordinate Reference System have been left out of the figure for clarity

3.3 The SPARQL-ST Query Language

This section presents the SPARQL-ST query language. We first give a formal syntax for SPARQL-ST and present a formal semantics for SPARQL-ST queries. We then illustrate the concrete syntax of SPARQL-ST with a series of examples. At the end of this section, we present motivations for various aspects of the SPARQL-ST design and discuss possible alternatives.

3.3.1   Formal Syntax for SPARQL-ST

In this section, we give a formalization of the SPARQL-ST syntax that is based on
the formalization of the SPARQL syntax given by [19]. We introduce spatial variables and temporal variables, which are used to form spatiotemporal graph patterns.
We also introduce spatial built-in conditions and temporal built-in conditions.

3.3.1.1   Spatiotemporal Graph Patterns

Let UL denote the union $U \cup L$ (recall that U is the set of URIs and L is the set of
Literals) and let V_N be a set of variables. Let V_S be a set of spatial variables, and
let V_T be a set of temporal variables. V_N, V_S, V_T, and RT (the set of RDF terms)
are pairwise disjoint. A *spatial triple pattern* is a 3-tuple from $(UL \cup V_N \cup V_S) \times$
$(U \cup V_N) \times (UL \cup V_N \cup V_S)$. A *spatiotemporal triple pattern* is a 4-tuple from $(UL \cup$
$V_N \cup V_S) \times (U \cup V_N) \times (UL \cup V_N \cup V_S) \times (V_T)$. A *spatiotemporal graph pattern* is
defined recursively as follows:

- If *st* is a spatial triple pattern, then *st* is a spatiotemporal graph pattern
- If *stt* is a spatiotemporal triple pattern, then *stt* is a spatiotemporal graph pattern
- If SP_1 and SP_2 are spatiotemporal graph patterns, then $(SP_1$ AND $SP_2)$ is a
  spatiotemporal graph pattern
- If SP is a spatiotemporal graph pattern and R is a SPARQL built-in condition,
  then the expression $(SP$ FILTER $R)$ is a spatiotemporal graph pattern
- If SP is a spatiotemporal graph pattern and SR is a spatial built-in condition, then
  the expression $(SP$ SPATIAL FILTER $SR)$ is a spatiotemporal graph pattern
- If SP is a spatiotemporal graph pattern and TR is a temporal built-in condition,
  then the expression $(SP$ TEMPORAL FILTER $TR)$ is a spatiotemporal graph
  pattern

The syntax for SPARQL built-in conditions is given in [19] and remains unchanged. Spatial built-in conditions and temporal built-in conditions are described
below.

3.3.1.2   Spatial Built-In Conditions

SPARQL-ST requires that we express spatial constraints on spatial variables. We
introduce spatial built-in conditions for this purpose. Spatial built-in conditions are
built from qualitative spatial expressions and metric spatial expressions.

A *qualitative spatial function* is a Boolean function $qsf : S \times S \rightarrow \mathbb{B}$, where S is
the set of all possible spatial geometries as defined by the ontology in Fig. 3.2. Any
of the following topological spatial relations identified by Egenhofer and Herring
[7] may be used as qualitative spatial functions in our formalization: *disjoint, touch,*

overlap boundary disjoint, overlap boundary intersect, equal, contains, covers, inside, covered by. We define a *qualitative spatial expression, qse*, as follows, where $s_1, s_2 \in S \cup V_S$.

$$\langle qse \rangle ::= qsf(s_1, s_2)$$

A *metric spatial function* is a function $msf : S \times S \rightarrow \mathbb{R}$. We use one metric spatial function $distance : S \times S \rightarrow \mathbb{R}$, which returns the distance between two spatial geometries. We define a *metric spatial expression, mse*, as follows, where $s_1, s_2 \in S \cup V_S$ and $r \in \mathbb{R}$.

$$\langle mse \rangle ::= \langle msf(s_1, s_2) \rangle \langle comp \rangle r$$

$$\langle comp \rangle ::= < \mid > \mid \leq \mid \geq \mid =$$

A *spatial built-in condition sf* evaluates to a Boolean value for a given graph and is defined in terms of metric spatial expressions and qualitative spatial expressions. A spatial built-in condition takes the following form.

$$\langle sf \rangle ::= \langle mse \rangle \mid \langle qse \rangle \mid \langle sf \rangle \textbf{ AND } \langle sf \rangle \mid \langle sf \rangle \textbf{ OR } \langle sf \rangle \mid \textbf{NOT } \langle sf \rangle$$

3.3.1.3 Temporal Built-In Conditions

To express constraints on temporal variables in SPARQL-ST, we introduce temporal built-in conditions. Temporal built-in conditions are built from qualitative and metric temporal expressions. For a given temporal RDF graph G_t over time domain T, let I denote the set of all time intervals over T.

As a prerequisite, we define a temporal primitive tp as follows, where $V_T' \subseteq V_T$, $vt \in V_T$ and $i \in I$.

$$\langle tp \rangle ::= intersect(V_T') \mid range(V_T') \mid vt \mid i$$

A *qualitative temporal function* is a Boolean function $qtf : I \times I \rightarrow \mathbb{B}$. Any of the thirteen interval relations identified by Allen [2] can be used in qualitative temporal functions in our formalization. We define a *qualitative temporal expression, qte*, as follows.

$$\langle qte \rangle ::= qtf(\langle tp \rangle, \langle tp \rangle)$$

A *metric temporal function* is a function $mtf : I \times I \rightarrow \mathbb{Z}$. We use one metric temporal function $elapsed\_time : I \times I \rightarrow \mathbb{Z}$, which is defined for two disjoint time intervals as the duration of time between the end of the earliest interval and the start of the latest interval. The function returns zero if the intervals are not disjoint. We define a *metric temporal expression, mte*, as follows, where $z \in \mathbb{Z}$.

$$\langle mte \rangle ::= \langle mtf(\langle tp \rangle, \langle tp \rangle) \rangle \langle comp \rangle z$$

$$\langle comp \rangle ::= < \mid > \mid \leq \mid \geq \mid =$$

A *temporal built-in condition* tf evaluates to a Boolean value for a given graph and is constructed from qualitative temporal expressions and metric temporal expressions as follows:

$$\langle tf \rangle ::= \langle mte \rangle \mid \langle qte \rangle \mid \langle tf \rangle \text{ AND } \langle tf \rangle \mid \langle tf \rangle \text{ OR } \langle tf \rangle \mid \text{ NOT } \langle tf \rangle$$

3.3.2   Formal Semantics for SPARQL-ST

We first give some initial definitions and then present the formal semantics of SPARQL-ST.

3.3.2.1   Initial Definitions

Let T be a set of totally ordered time points. Let G_t be a temporal RDF graph defined over T. $TRIPLES(G_t)$ denotes the set $\{(s,p,o) \mid \exists t \in T \text{ with } (s,p,o) : [t] \in G_t\}$. For each statement $e = (s,p,o) \in TRIPLES(G_t)$, let $temporal(e) = \{t \mid (s,p,o) : [t] \in G_t\}$. For a set of time points $T' \subseteq T$, let $contig\_intervals(T') = \{[t_i,t_j] \mid \forall t \in T : (\text{if } t_i \leq t \text{ and } t \leq t_j \text{ then } t \in T') \text{ and } t_{i-1} \notin T' \text{ and } t_{j+1} \notin T'\}$. Consider the following example: suppose $T = \{1,2,3,4,5,6,7,8,9,10\}$ and $T' = \{2,3,4,7,8\}$, then $contig\_intervals(T') = \{[2,4],[7,8]\}$.

Given a set of time intervals $I = \{(s_1,t_1), (s_2,t_2), ..., (s_n,t_n)\}$ defined over T, let $s_{min} = \min_{1 \leq i \leq n} s_i$, $s_{max} = \max_{1 \leq i \leq n} s_i$, $t_{min} = \min_{1 \leq i \leq n} t_i$, and $t_{max} = \max_{1 \leq i \leq n} t_i$. We define two values, *intersect* and *range*, as follows: $intersect(I) = [s_{max}, t_{min}]$ if $s_{max} \leq t_{min}$, else *null*, $range(I) = [s_{min}, t_{max}]$ if $s_{min} \leq t_{max}$, else *null*. Conceptually, $intersect(I)$ is the largest time interval that intersects each interval in I, and $range(I)$ is the smallest interval that contains each interval in I.

3.3.2.2   SPARQL-ST Semantics

The semantics of a SPARQL-ST spatiotemporal graph pattern query are based on the concept of a mapping introduced by Perez et al. in [19] to provide a formal semantics for SPARQL. Here, we extend this mapping concept to also include spatial and temporal variables. Conceptually, our extension maps spatial variables to a set of RDF triples rather than a single URI and maps temporal variables to a time interval rather than a single URI. Recall that for a set A, 2^A denotes the powerset of A. A mapping μ is a function from $(V_N \cup V_S \cup V_T)$ to $(RT \cup 2^{((U \cup B) \times U \times RT)} \cup I)$ such that:

- If $vn \in V_N$ then $\mu(vn) = rt \in RT$
- If $vs \in V_S$ then $\mu(vs) = g \in 2^{((U \cup B) \times U \times RT)}$ and g forms a valid Geometry instance
- If $vt \in V_T$ then $\mu(vt) = i \in I$

```
@prefix rdf: <http://www.w3.org/1999/02/22-rdf-syntax-ns#> .
@prefix rdfs: <http://www.w3.org/2000/01/rdf-schema#> .
@prefix geo: <http://knoesis.org/geo#> .

geo:polygon_123 rdf:type geo:Geometry .
geo:polygon_123 rdf:type geo:GeometricPrimitive .
geo:polygon_123 rdf:type geo:Polygon .
geo:polygon_123 geo:exterior geo:LinearRing_123 .
geo:LinearRing_123 geo:lrPosList
                         "-122.84501 42.240328, -122.8075 42.240328,
                          -122.8075 42.3764, -122.84501 42.3764,
                          -122.84501 42.240328" .
geo:polygon_123 geo:uses_crs geo:CRS_NAD83 .
geo:CRS_NAD83 geo:srsName "urn:ogc:def:crs:EPSG:6.6:4269" .
```

Fig. 3.3 Set of triples representing a polygon

For a mapping μ, the subset of $(V_N \cup V_S \cup V_T)$ where it is defined is called its domain $dom(\mu)$. Two mappings μ_1 and μ_2 are compatible if, for all $x \in dom(\mu_1) \cap dom(\mu_2)$, it is the case that $\mu_1(x) = \mu_2(x)$. In other words, the union $\mu_1 \cup \mu_2$ is also a mapping. In addition, for two sets of mappings M_1 and M_2, the *join* is defined as:

$$M_1 \bowtie M_2 = \{\mu_1 \cup \mu_2 \mid \mu_1 \in M_1 \text{ and } \mu_2 \in M_2$$
$$\text{and } \mu_1 \text{ and } \mu_2 \text{ are compatible mappings}\}$$

The semantics of a spatiotemporal graph pattern are defined in terms of a function $[[\cdot]]$, which takes a spatiotemporal graph pattern and returns a set of mappings. Before we can define this function, we need to introduce some additional constructs to handle spatial and temporal aspects of graph patterns. Because a spatial variable maps to a collection of triples, we introduce a function, *head*, that reduces this set of triples to a single URI. We also define functions, *triple* and *t_triple*, which allow us to go from a mapping to a single RDF triple or temporal RDF triple. These single triples are used to formally define the function $[[\cdot]]$.

We will first define the function $head : (RT \cup 2^{((U \cup B) \times U \times RT)}) \rightarrow RT$. This function is defined as follows:

- If $t \in RT$ then $head(t) = t$
- If $t \in 2^{((U \cup B) \times U \times RT)}$ then $head(t) = s \in RT$ such that $(s, rdf:type, Geometry) \in t$

Conceptually, if t is a single URI, $head(t)$ returns this single URI, and if t is a collection of triples representing a Geometry instance, $head(t)$ returns the top level URI of the Geometry instance. For the example in Fig. 3.3, the top level URI is $geo : polygon\_123$.

For a mapping μ and a spatial triple pattern sp, we denote the triple obtained by replacing the variables v in sp with the value $head(\mu(v))$ as $triple(\mu, sp)$. For a mapping μ and a spatiotemporal triple pattern stp, we denote the temporal triple obtained by replacing the variables $v \in V_N \cup V_S$ in stp with the value $head(\mu(v))$ and the variables $t \in V_T$ in stp with the value $\mu(t)$ as $t\_triple(\mu, stp)$.

Let G_t be a temporal RDF graph, sp a spatial triple pattern, stp a spatiotemporal triple pattern and SP_1, SP_2 spatiotemporal graph patterns. The evaluation of a spatiotemporal graph pattern over G_t, denoted $[[\cdot]]_{G_t}$, is defined recursively as:

- $[[sp]]_{G_t} = \{\mu \mid dom(\mu) = var(sp)$ and
  $\qquad triple(\mu, sp) \in TRIPLES(G_t)\}$
- $[[stp]]_{G_t} = \{\mu \mid dom(\mu) = var(sp)$ and for
  $\qquad (s, p, o) : [t_1, t_2] = t\_triple(\mu, stp)$
  $\qquad$ it is the case that $(s, p, o) \in TRIPLES(G_t)$ and
  $\qquad [t_1, t_2] \in contig\_intervals(temporal((s, p, o)))\}$
- $[[SP_1 \text{ AND } SP_2]]_{G_t} = [[SP_1]]_{G_t} \bowtie [[SP_2]]_{G_t}$

The semantics of spatial built-in conditions and temporal built-in conditions are defined as follows. A mapping μ satisfies a spatial built-in condition sf written $\mu \models sf$ if $var(sf) \subseteq dom(\mu)$ and sf evaluates to true when each variable $vs \in V_S$ in sf is replaced with $geom(\mu(vs))$. Note that the function $geom : 2^{((U \cup B) \times U \times RT)} \rightarrow \mathbb{R}^2$ maps the RDF serialization of a Geometry to an actual point, line or polygon. A mapping μ satisfies a temporal built-in condition tf written $\mu \models tf$ if $var(tf) \subseteq dom(\mu)$ and tf evaluates to true when each variable $vt \in V_T$ in tf is replaced with $\mu(vt)$.

Given a temporal RDF graph G_t, a spatiotemporal graph pattern SP, a spatial built-in condition SR and a temporal built-in condition TR,

- $[[SP \text{ SPATIAL FILTER } SR]]_{G_t} = \{\mu \in [[SP]]_{G_t} \mid$
  $\qquad\qquad\qquad \mu \models SR\}$
- $[[SP \text{ TEMPORAL FILTER } TR]]_{G_t} = \{\mu \in [[SP]]_{G_t} \mid$
  $\qquad\qquad\qquad \mu \models TR\}$

3.3.3   SPARQL-ST by Example

This section presents the concrete syntax of SPARQL-ST using examples. Temporal variables are identified using a '#' prefix, and spatial variables are identified using a '%' prefix. The constructs *intersect()* and *range()* refer to the *intersect* and *range* intervals defined in Sect. 3.3.2.2.

(Temporal Filter Query) Find all house members who sponsored a bill after April 2, 2008. This query returns each representative and the *intersect* interval representing the time the bill was sponsored. This query uses the *TEMPORALFILTER* construct to ensure that the bill was sponsored after April 2, 2008.

```
SELECT ?p, intersect(#t1, #t2, #t3, #t4)
    WHERE {
    ?p usgov:hasRole ?r #t1 .
    ?r usgov:forOffice ?o #t2 .
```

```
?o usgov:isPartOf usgov:congress/house #t3 .
?p usgov:sponsor ?b #t4 .
TEMPORAL FILTER
(
after(intersect(#t1, #t2, #t3, #t4),
      interval(04:02:2008, 04:02:2008,
      MM:DD:YYYY)))}}
```

(Basic Spatial Query) Find the congressional district spatial geometries for all politicians who voted "Aye" for bill number 88. This query simply selects the spatial variable representing the appropriate Geometry instance.

```
SELECT ?p, %g
    WHERE {
           ?v usgov:hasBallot ?b .
           ?v usgov:billNo "88" .
           ?b usgov:voter ?p .
           ?b usgov:hasOption "Aye" .
           ?p usgov:hasRole ?r .
           ?r usgov:forOffice ?o .
           ?o usgov:represents ?c .
           ?c stt:located_at %g }}
```

(Filtered Spatiotemporal Query) Find all politicians representing congressional districts within a given bounding box and return the times that those politicians represented those areas. This query uses a *SPATIALFILTER* involving the *inside* function to ensure each returned congressional district falls within the given geographical area. The intersect interval of several temporal variables is used to select the desired temporal intervals.

```
SELECT ?p, intersect(#t1, #t2, #t3, #t4)
    WHERE {
           ?p usgov:hasRole ?r #t1 .
           ?r usgov:forOffice ?o #t2 .
           ?o usgov:represents ?c #t3 .
           ?c stt:located_at %g #t4 .
           SPATIAL FILTER (inside(%g, GEOM(POLYGON ((
           -75.14 40.88, -70.77 40.88, -70.77 42.35,
           -70.77 42.35, -75.14 42.35,
           -75.14 42.35, -75.14 40.88)) ))}}
```

3.3.4   Design Decisions

The introduction of spatial variables is a major component of our SPARQL extension. These variables represent complex spatial objects and map to a set of RDF triples. Two possible alternatives to introducing a new variable type are: (1) specifying all parts of the spatial object in a graph pattern and (2) utilizing the concept of named graphs to represent spatial objects.

The example below illustrates the first alternative where the relevant parts of a spatial object are specified in a graph pattern.

```
SELECT ?positions
WHERE {
<http://house/106/nh> usgov:represents ?x .
?x stt:located\_at ?sr .
?sr geo:exterior ?lr .
?lr geo:lrPosList ?positions }
```

We see the following problems with this approach. First, the relevant portions of a spatial object that need to be returned from the query will vary. For example, if one is selecting the position lists of a multipolygon, it is unclear how to specify this in a graph pattern, as the number of polygons making up each multipolygon will vary. Second, it is unclear how to reference a spatial object in a spatial filter expression. That is, what parts of the graph pattern should be passed to a spatial function in the spatial filter expression? A special variable type solves both of these problems.

Another alternative is to use named graphs to represent spatial objects. A named graph is created by associating a set of RDF triples with some URI u. This set of triples can then be collectively referred to by the identifier u. A query using this approach is shown in the example below. This query returns all triples making up each named graph (geometry) in the result.

```
SELECT ?g, ?s, ?p, ?o
WHERE {
<http://house/106/nh> usgov:represents ?x .
?x stt:located_at ?g .
GRAPH ?g {?s, ?p, ?o} }
```

We feel that this approach makes the semantics of our STT modeling approach less clear because it hides the fact that the query is dealing with spatial objects. In addition, using a named graph as input to a spatial function could lead to unexpected errors if the input named graph did not represent a spatial geometry.

Another key aspect of our approach is using temporal variables to specify a quad to represent a temporal triple pattern. An alternative would be to use SPARQL as it is and use the RDF reification triples to extract valid times for triples. This approach is problematic for the following reasons. First, it is extremely verbose, as it would take eight triple patterns to retrieve the valid times for each statement. Second, the semantics of temporal RDF are lost because the query will simply match triples in the RDF dataset, and the concepts of temporal RDFS inferencing (see Sect. 3.4.1) are lost. In addition, special temporal variables make it clear that one is querying a temporal RDF graph rather than a plain RDF graph.

3.4 Implementation Framework

We have implemented SPARQL-ST by extending a commercial relational database that supports spatial objects.[3] We provide a single SQL table function, *sparql_st*, that inputs a valid SPARQL-ST query and returns a table of the resulting variable mappings. Our prototype implementation supports qualitative spatial and temporal relationships and spatial and temporal filter expressions involving conjunctions of filtering conditions.

3.4.1 Storage and Indexing Scheme

Our storage scheme for spatiotemporal RDF is shown in Fig. 3.4. RDF triples are stored using the schema-oblivious storage scheme [28]. A *URIID* table maps full URIs to numeric ids, and a *Triples* table stores subject, predicate and object ids. This basic scheme is augmented with additional structures for efficient processing of spatial and temporal data. A *TemporalTriples* table stores subject, predicate and object ids along with two datetime columns that represent the start and end of the triple's valid time interval. A *SpatialData* table maps Geometry URIs with their representation in the native spatial object type of the database. This table also stores the RDF/XML serialization of the Geometry (e.g., the triples in Fig. 3.3) to

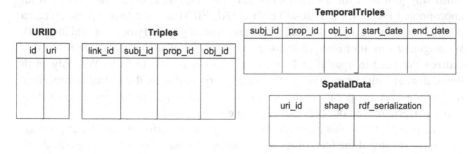

Fig. 3.4 Table structures used for our SPARQL-ST implementation

---

[3]License restrictions related to publication of timing results prevent us from disclosing the name of the database vendor.

allow for efficient retrieval of spatial variable mappings. The *TemporalTriples* and *SpatialData* tables are constructed during a post processing step after all asserted triples are loaded into the *URIID* and *Triples* tables.

The complete set of asserted and inferred temporal triples is stored in the *TemporalTriples* table. A post processing step performs RDF/S inferencing and computes the valid time intervals for inferred statements. For example, given the asserted temporal triples $(x, p, y) : [1,5], (p, rdfs:domain, a) : [0,10]$, we would infer $(x, rdf:type, a) : [1,5]$ through rule *rdfs2* (refer to [10] for the complete set of RDFS inferencing rules). In each case, the computed valid time interval is the intersection of the valid time intervals of the set of triples used to make the inference. Temporal evolution of ontology schemas is beyond the scope of this work, so we therefore limit temporal RDFS inferencing to instance-level statements. That is, we assume schema-level statements are valid during the interval $[0, \infty]$, and we compute valid times for all *rdf:type* statements inferred through rules *rdfs2*, *rdfs3* and *rdfs9* and all instance-level statements inferred from *rdfs:subPropertyOf* (i.e. rule *rdfs7*). We ensure that the final valid times recorded for each statement are stored as the minimal set of contiguous intervals as described in Sect. 3.3.2.1. The algorithm for this post processing step is given in our earlier work [22].

3.4.2   Query Evaluation Procedure

The evaluation of a SPARQL-ST query proceeds in two basic steps. First, the SPARQL-ST query is translated into a SQL query against the table structures described in Sect. 3.4.1. This initial query is referred to as the base query. Second, further processing of the results of the base query is done on a row-by-row basis, and the appropriate result set is constructed and returned.

The first step in our query evaluation procedure is construction of the base SQL query for a given SPARQL-ST query. We first translate the graph pattern into a multi-way join over the *TemporalTriples*, *URIID*, and *SpatialData* tables. If an appropriate SPATIAL FILTER or TEMPORAL FILTER condition is present (i.e., a condition involving a variable and a constant spatial geometry or temporal interval), we augment this multi-way join query with a spatial or temporal predicate that utilizes the built-in spatial and temporal indexes of the DBMS. We only push down a single filter condition to the base query, and spatial conditions are given preference over temporal conditions. Spatial conditions are favored due to their better performance in our previous experiments [22].

The second step in our query evaluation procedure performs additional processing on the results of the base query on a row-by-row basis. In this step, we evaluate any filter conditions that were not pushed down to the base query, and we construct any *intersect* or *range* intervals that need to be returned from the table function. We also construct and return a result row of the table function in this step.

3.5   Performance Evaluation

The experimental evaluation of our implementation is described in this section. All experiments were run on a Sun Fire V490 server with four 1.8 GHz Ultra Sparc IV processors and 8GB of main memory. The operating system used was 64-bit Solaris 9. The database used an 8 KB block size and was configured with a 512 MB buffer cache and a sort area size of 512 MB. The times reported for each query were obtained as follows. The query was run once initially to warm up the database buffers and then timed for five consecutive executions. We report the mean execution time over these five consecutive executions.

For testing, B^+-Tree indexes were created on each column of the *TemporalTriples* table and on the *value_id* column of the *SpatialData* table, and an R-Tree index was created on the *shape* column of *SpatialData*. We also created four composite B^+-Tree indexes on the *TemporalTriples* table to allow for efficient index-based joins: (*prop_id, subj_id, obj_id*) and (*prop_id, obj_id, subj_id*) for spatial operators and (*prop_id, subj_id, obj_id, start, end*) and (*prop_id, obj_id, subj_id, start, end*) for temporal operators.

Testing details (e.g., queries used and datasets) are available at.[4]

3.5.1   Datasets

We conducted experiments using two RDF datasets. One consisted of synthetically generated RDF data corresponding to historical analysis of WWII (SynHist), and the other (GovTrack) consisted of real-world RDF data from the political domain that we obtained from http://www.govtrack.us/data/rdf/. Table 3.1 shows the characteristics of these datasets.

Table 3.1 Characteristics of GovTrack and SynHist datasets

Dataset	Num triples (in thousands)			Num spatial features	Avg num points per polygon
	Asserted	Inferred	Total		
SH1	71	50	121	3,470	98
SH2	980	643	1,623	28,488	63
SH3	4,295	2,708	7,002	77,440	67
SH4	11,593	7,559	19,152	169,722	56
SH5	17,616	11,290	28,906	244,653	61
GT1	2,959	3,036	5,995	3,433	4
GT2	5,245	5,226	10,471	3,433	4
GT3	12,820	13,099	25,919	3,433	4

---

[4]http://knoesis.wright.edu/students/mperry/sparql-st.html.

3.5.1.1   SynHist Dataset

Five synthetically generated datasets (SH1–SH5) were used in our experiments.
The datasets correspond to a historical battlefield analysis ontology schema that
we created. The ontology schema defined 15 class types and 9 property types.
Each dataset was created in three phases. First we populated the thematic portion
of the ontology. Second we added spatial information, and in the final step we
generated temporal labels for the statements in the populated ontology. To populate
the thematic portion of the battlefield analysis ontology, we used the ontology
population tool described in [20]. This tool inputs an ontology schema and relative
probabilities for generating instances of each class and property type. Based on these
probabilities, it generates instance data, which, in effect, simulates the population of
the ontology.

To add spatial aspects to this dataset, we randomly assigned a spatial geometry
to each instance of Geometry in the ontology. We used year 2000 census block
group boundary polygons from the US Census Bureau[5] for the spatial geometries.
Differently-sized sets of contiguous US States were chosen in proportion with the
ontology size.

The final phase of dataset generation assigned temporal labels to statements in
the ontology. Temporal intervals were randomly assigned to each asserted instance
statement. Start times and end times for each interval were randomly selected with
uniform probability from two overlapping date ranges. We ensured that each interval
was valid (i.e., start time earlier than end time) before adding it to the dataset.

3.5.1.2   GovTrack Dataset

The GovTrack RDF dataset contains data about activities of the US Congress. More
specifically, it contains data describing politicians, bills, voting records, political
organizations, political offices, and terms held by politicians. The ontologies used
for this dataset contained 74 classes and 139 properties. 22 classes and 47 properties
were actually used in the instance data. Some transformations and enhancements
of the dataset were needed to make it appropriate for experimentation. The
GovTrack data contained a significant amount of temporal information. However,
this information was encoded using separate properties rather than as temporal
RDF. A preprocessing step was therefore needed to transform the dataset into a
temporal RDF graph. To enhance the dataset with spatial data, we linked *Congres-
sional_District* instances with a bounding box representation of their corresponding
boundary polygons available from the US Census Bureau. We used boundary files
for the 106th–110th Congress. We created three differently-sized subsets of the
GovTrack data (GT1–GT3). GT1 contained information on bills and voting from
the 106th Congress. GT2 used the 106th and 107th Congress, and GT3 used the
106th–110th Congress.

---

[5]http://www.census.gov/geo/www/cob/bdy_files.html.

3.5.2 Experiments

Our experiments were designed to characterize the overall performance of our approach with respect to (1) dataset size and (2) graph pattern complexity.

In the following, we refer to two different graph pattern types: unselective and selective. An *unselective graph pattern* contains constant URIs in the predicate position in each triple pattern and variables in each subject and object position, for example:

```
?x usgov:cosponsor ?y .
?x usgov:sponsor ?z .
?x usgov:inCommittee ?c
```

A *selective graph pattern* has constant URIs in each predicate position and additionally contains a constant URI in the subject and/or object position in at least one triple pattern, for example:

```
?p usgov:hasRole ?y .
?y usgov:forOffice usgov:congress/senate/va
```

3.5.2.1 Scalability w.r.t. Dataset Size

Tables 3.2 and 3.3 summarize the results of our experiments with respect to dataset size. These experiments were designed to test the scalability of our implementation for a basic set of SPARQL-ST queries.

Temporal Selection: Queries G1 and H1 select the intersect interval of the triples making up 5 hop selective graph patterns. The results show that query execution time is near constant as the dataset size grows. This is a result of the index-based nested loop join (NLJ) strategy used by the DBMS, which tends to have execution times proportional to the result set size.

Temporal Filter: Queries G2 and H2 test the scalability of our implementation for a SPARQL-ST query involving a TEMPORAL FILTER condition between a derived intersect interval and a constant time interval. These queries used an unselective graph pattern in combination with very selective temporal conditions. The queries show relatively constant execution time for the GovTrack dataset but show more of a linear growth for the SynHist dataset. In each case, the DBMS uses an index-based NLJ strategy over the composite indexes containing start date and end date information.

These particular queries represent a challenging case for our implementation. Because the INTERSECT/RANGE interval derived for a graph pattern instance is constructed dynamically from the temporal labels of edges in the graph pattern instance, we cannot directly index these derived values. We must instead apply the temporal filtering condition to each graph pattern instance as it is being constructed,

Table 3.2 Scalability with respect to dataset size for GovTrack dataset. Legends: # T = Number of triples, # V = Number of variables, # R = Result size

Query	Description	# T	# V	# R	Execution time (s)		
					GT1	GT2	GT3
G1	t-select	5	5	94	0.14	0.136	0.137
G2	t-filter – int / after	5	6	483	0.614	0.609	0.565
G3	t-join – int / during	3/3	3/3	90	0.821	0.817	0.838
G4	t-join – int / before	3/3	3/3	120	0.376	0.376	0.375
G5	s-select	5	5	428	2.663	2.658	2.660
G6	s-filter – anyinteract	5	6	562	3.340	3.360	3.345
G7	s-join – overlap	4/1	4/2	144	0.99	0.995	0.981
G8	s-filter – anyinteract	5	6	397	3.444	3.438	3.463
	+ t-filter – int / during						

Table 3.3 Scalability with respect to dataset size for SynHist dataset. Legends: # T = Number of triples, # V = Number of variables, # R = Result size

Query	Description	# T	# V	# R	Execution time (s)				
					SH1	SH2	SH3	SH4	SH5
H1	t-select	5	5	178	0.290	0.291	0.290	0.290	0.292
H2	t-filter – int / overlap	5	6	128	0.178	0.321	0.572	1.179	2.238
H3	t-join – int / overlap	3/3	3/3	184	0.808	0.838	0.896	1.020	1.108
H4	t-join – int / anyinteract	3/3	3/3	42	0.360	0.361	0.374	0.389	0.392
H5	s-select	5	5	224	3.267	3.266	3.258	3.256	3.275
H6	s-filter – inside	5	6	303	3.999	3.996	3.984	3.989	3.981
H7	s-join – equal	4/2	4/2	48	0.296	0.296	0.298	0.297	0.295
H8	s-filter – inside	5	6	13	3.772	3.770	3.779	3.768	3.781
	+ t-filter – int / overlap								

which can lead to a very large set of intermediate results that are later discarded. The unnecessary intermediate results are generated because, in many cases, we cannot exclude a graph pattern instance until it is fully constructed and the final derived time interval is known. We try to alleviate this problem by placing limited temporal constraints on individual triple patterns in the graph pattern. These initial constraints can reduce the number of intermediate results generated, but the amount of reduction depends on the specific interval type and temporal relation used. This issue is further explored in Sect. 3.5.2.2.

The difference in the scalability of the queries over the GovTrack dataset is a result of the characteristics of the time intervals in each dataset. The triples in the SynHist dataset have more densely packed valid time intervals with a higher degree of overlap than do the triples in the GovTrack dataset. As a result, the temporal filtering conditions that can be placed on each triple in the graph pattern are ultimately less selective, leading to larger growth in intermediate results as the dataset size increases.

Temporal Join: Queries G3, G4 and H3, H4 tested the scalability of our implementation for SPARQL-ST queries involving a TEMPORAL FILTER condition between two derived time intervals. The filter condition acts as a join between two disjoint graph patterns. The execution times for queries G3, G4 and H4 are relatively constant as the dataset size grows, but query H3 shows a slight growth in execution time. The growth for this query results from a combination of the particular temporal relation used and the denser set of time intervals in the SynHist dataset.

Spatial Selection: Queries G5 and H5 select a spatial variable. These queries use a selective graph pattern involving a single spatial variable. As a result of the index-based join strategy used by the DMBS, query execution time is near constant as dataset size increases. These queries have a significantly longer execution time than the corresponding temporal selection queries. The longer time is a result of the overhead of populating the result set of the query with RDF/XML serialization (stored as a CLOB) of each spatial feature in the result.

Spatial Filter: Queries G6 and H6 use an unselective graph pattern in combination with a SPATIAL FILTER expression over a spatial variable and a constant spatial feature (a rectangle in each case). The execution times for each query are near constant as the dataset size increases due to the index-based join strategy used by the DBMS to evaluate the graph pattern. The execution times for the GovTrack dataset are a bit faster because the spatial features in this dataset are simpler. Again, extra time is needed for spatial queries to populate the result set.

In the SynHist dataset, we see that the spatial filtering queries scale better than temporal filtering queries. Unlike INTERSECT/RANGE intervals, the spatial geometries can be indexed because they are not dynamically created. The spatial filtering queries consequently scale better because we can consistently reduce the search space using the spatial index and do not get as much growth in intermediate results as the dataset size increases.

Spatial Join: Queries G7 and H7 involved a graph pattern with two disjoint components and a SPATIAL FILTER condition over two spatial variables that acts as a spatial join for the two components of the graph pattern. Again, the execution times for each query are near constant as the dataset size increases as a result of the index-based join strategy used to evaluate the graph pattern. The times are a bit faster than other spatial queries because of the smaller result set sizes for these queries, which limits the overhead of populating the result set.

Spatiotemporal Filter: Queries G8 and H8 involve unselective graph patterns and both SPATIAL FILTER and TEMPORAL FILTER conditions. In each case, query execution time is near constant as the dataset size increases. Queries over the SynHist dataset are slower relative to their result set size because of the less efficient temporal processing and more complicated spatial features in this dataset.

3.5.2.2 Scalability w.r.t. Graph Pattern Complexity

Our next experiments are designed to test the scalability of our implementation with respect to query complexity: that is, the size of the graph pattern used. All experiments used the GT3 and SH5 datasets.

Temporal Filter: Our first experiment tested TEMPORAL FILTER queries involving unselective graph patterns and selective temporal filtering conditions. The key to the performance of these queries is to reduce the amount of search space by placing partial temporal constraints on individual triple patterns in the graph pattern. As we noted earlier, the effectiveness of these partial temporal constraints depends on the particular interval type and temporal relation used in a query.

The objective of this experiment was to characterize the performance of temporal filter queries in both the worst-case scenario (very limited initial temporal filtering) and the best-case scenario (complete initial temporal filtering). An INTERSECT interval type in combination with a DURING temporal relation represented the worst-case. In this situation, we can only enforce that the valid time interval of each triple does not end before the query interval starts or start after the query interval ends. In contrast, with a RANGE interval type and a DURING temporal relation, we can enforce that each triple starts after the query interval starts and ends before the query interval ends. These conditions completely filter out any unwanted graph pattern instances, and this query represents a best-case. Figure 3.5a shows the execution times for a best-case and worst-case query for unselective graph patterns varying in size from one triple to seven triples. We can see that execution time grows roughly linearly in each case, but performance is significantly worse for the worst-case scenario. The performance is better for the GovTrack dataset because of the nature of the temporal intervals in each dataset as we discussed in Sect. 3.5.2.1. The execution time for queries over the SynHist dataset tends to grow more rapidly at first and then taper off as the graph pattern gets more complex. This trend is a result of the selectivity of the graph pattern itself. In this dataset, there are fewer instances of the more complex graph patterns. This slows the growth in intermediate results, so not as much additional temporal filtering is needed after executing the base query.

Spatial Filter: Our next experiment tested SPATIAL FILTER queries involving unselective graph patterns and selective spatial filtering conditions. Figure 3.5b shows the execution times of these queries. As the graph pattern size grows, the query execution times show linear scalability on both datasets. These spatial queries are initially slower than the temporal filter queries but become faster for the larger graph patterns because the time for temporal filtering outweighs the time needed to populate spatial features in the result set. The faster execution times result from the more effective spatial indexing. The spatial index is initially used to select the URIs satisfying the spatial filtering condition, which reduces the search space for evaluating the rest of the graph pattern. The queries over the SynHist dataset have slower execution times because spatial computations are more expensive for the more complex spatial geometries in the SynHist dataset.

Basic Selection Queries: Our final experiments tested the scalability of our implementation for spatial, temporal and spatiotemporal selection queries using selective graph patterns ranging in size from 1 to 10 triples. The results of these experiments are shown in Fig. 3.5c–h. The number of result rows returned from the query is also shown in the graphs. These graphs show that performance is quite good

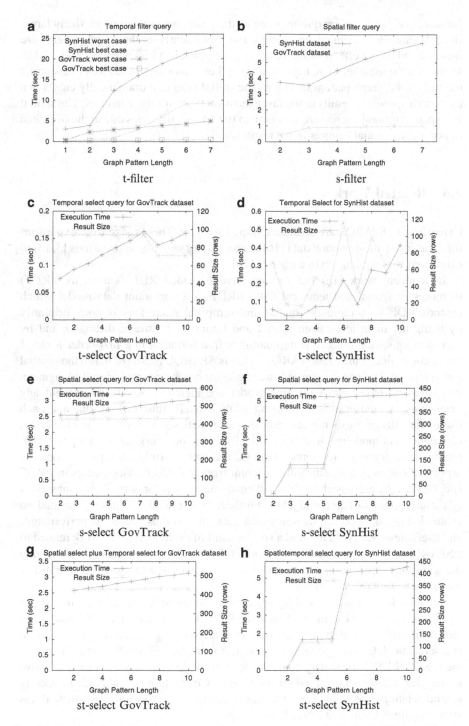

Fig. 3.5 Results of scalability experiments for graph pattern complexity for SynHist (SH5) and GovTrack (GT3) datasets

for selective graph pattern queries even as the graph patterns grow relatively large. In each case, the execution times grow roughly linearly as the graph pattern size increases when the effects of the result set size are taken into account. The DBMS starts with the most selective triple pattern and uses an index-based join to construct the rest of the graph pattern instance. The initial selection dramatically cuts down the search space and results in the fast execution times for these queries. The spatial and spatiotemporal queries are slower than the temporal queries due to the overhead of populating spatial features in the result set.

3.6 Related Work

Extensions of SPARQL are abundant in the literature. These range from extensions for handling spatio-temporal data [16] to computing semantic associations [3, 13] to extensions for enabling skyline queries [25].

In a recent work [16] the authors have extended RDF (known as stRDF) to represent spatial and temporal data. stRDF is a constraint data model which extends RDF with represent spatial and temporal data. This is done primarily by using the main ideas from spatial and temporal constraint databases and by representing spatial objects using quantifier-free formulas in a first-order logic of linear constraints. Further, stSPARQL extends SPARQL (stSPARQL) so that spatial and temporal data can be queried using a declarative and user-friendly language. Although the objective of the two works are identical, we do so without any extensions of modification to existing RDF Models, thus making our approach useful for querying existing real world spatial data resources such as GovTrack, Geonames. For modeling time, both stSPARQL and our work rely on temporal RDF graphs presented in [9] to represent the valid time of a triple. The spatial modeling aspects of our work is significantly different from stSPARQL. Geometries in stRDF and stSPARQL are based on the mathematical concept of semi-linear subsets of Q^k, using notions of linear algebra. Further, in our work we have presented an evaluation of our approach on real world datasets and thus proved its performance and usefulness. We could not find a similar kind of evaluation in the work related to stSPARQL. Thus to summarize, the works follow different approaches for reaching the same objective of supporting modeling and querying of spatio-temporal-data using Semantic Web technologies.

Another discussion of querying spatial data using SPARQL appears in a paper by Kolas and Self [15] in the Semantic Web in use track of ISWC 2007. The authors describe a prototype system for integrated storage and querying of spatial and semantic data. The system is queried using standard SPARQL syntax. They use the GeoRSS RDF vocabulary to model spatial objects and use a set of qualitative topological relationships based on the Region Connection Calculus [5] to specify spatial relationships in queries. The query below uses their approach to find gas stations within 1 mile of 38°N, 77°W.

```
SELECT ?x
WHERE {
?x rdf:type gas:GasStation .
?x georss:where ?y .
?y rcc:part ?p .
?y rcc:part ?p .
?p rdf:type gml:Buffer .
?p gml:radius "1" .
?p gml:bufferGeometry ?g .
?g rdf:type gml:Point .
?g gml:pos "38 -77" }
```

In contrast to this approach, we introduce special spatial variables and specify spatial constraints using a *SPATIALFILTER* clause instead of encoding the spatial constraint within the graph pattern. Without introducing spatial variables this approach would suffer from the shortcomings described in Sect. 3.3.4. In addition, their implementation only supported the relations *connected* and *part*, and no performance results were presented.

There have also been proposals for adding geospatial capabilities to SPARQL using the extensibility features of the Jena Semantic Web framework and its ARQ SPARQL engine [11]. For example, code implementing property functions that extend ARQ for geospatial relations appears at. [6]

The following example query uses a *nearby*() property function to select hotels near a certain point.

```
SELECT ?n
WHERE {
?s geo:nearby(51.45, -2.583) .
?s rdf:type ex:Hotel .
?s ex:name ?n}
```

Again, such an approach does not use spatial variables, so it will suffer from the shortcomings we mentioned earlier. In addition, property functions are an ARQ-specific feature that are not part of the SPARQL specification.

There are currently no extensions of SPARQL for temporal RDF graphs. However [8, 9, 24] discuss aspects of querying temporal RDF graphs. Gutierrez et al. [8, 9] briefly present a query language for temporal RDF graphs through a series of examples. The authors state that the query language needs a built in arithmetic language to reason about time and intervals and a construct to form maximal validity intervals for a given triple. In our proposal, the *TEMPORALFILTER* clause provides the needed temporal reasoning capabilities, and the need for maximal intervals is taken care of during our temporal RDFS inferencing procedure. Pugliese et al. formally define a temporal RDF query [24]. The query is essentially a

---

graph pattern involving triple patterns associated with either a temporal variable or a temporal constraint. The temporal query specified by Pugliese et al. also supports the notion of a maximal interval for each triple. An additional feature we support over these proposals is the ability to perform temporal computations over temporal intervals derived from the maximal intervals of multiple triples. We use the notions of *intersect* and *range* to provide this capability. Furthermore, neither of these works discuss extensions of SPARQL needed to support their proposed querying approaches.

3.7   Conclusions

This work presented SPARQL-ST, an extension of SPARQL for spatiotemporal queries. SPARQL-ST adds spatial variables and constructs for manipulating temporal triples. We gave a formal syntax and semantics for SPARQL-ST and presented a prototype implementation built on top of a commercial DBMS. We demonstrated the scalability of our prototype implementation with an experimental evaluation using both real-world and synthetic RDF datasets of over 25 million triples. In the future, we plan to investigate standardization issues with respect to our spatiotemporal extensions to SPARQL. We also plan to do a comparative study of the RDF/SPARQL approach for spatiotemporal querying presented in this work and an approach using OWL-DL and specialized spatial and temporal reasoners. Such a study would help determine the pros and cons of each method.

Acknowledgements We thank Professor T. K. Prasad for his helpful comments on our formalization of SPARQL-ST, and Cory Henson for his comments on a draft of this work. This work is partially funded by NSF-ITRIDM Award #0714441 (SemDIS: Discovering Complex Relationships in the Semantic Web) and by NSF Award #IIS-0842129, titled "III-SGER: Spatio-Temporal-Thematic Queries of Semantic Web Data: a Study of Expressivity and Efficiency (09/01/2008-08/31/2010)".

References

1. Alia I. Abdelmonty, Philip D. Smart, Christopher B. Jones, Gaihua Fu, and David Finch. A critical evaluation of ontology languages for geographic information retrieval on the internet. *Journal of Visual Languages and Computing*, 16(4):331–358, 2005.
2. James F Allen. Maintaining knowledge about temporal intervals. *Communications of the ACM*, 26(11):832–843, 1983.
3. Kemafor Anyanwu, Angela Maduko, and Amit P. Sheth. SPARQ2L: Towards support for subgraph extraction queries in RDF databases. In *16th International World Wide Web Conference*, pages 797–806, Banff, Alberta, Canada, 2007.
4. Dan Brickley and Ramanathan V. Guha. RDF vocabulary description language 1.0: RDF schema. W3C recommendation. http://www.w3.org/tr/rdf-schema/.

5. Anthony G Cohn, Brandon Bennett, John Gooday, and Nicholas Mark Gotts. Qualitative spatial representation and reasoning with the region connection calculus. *GeoInformatica*, 1(3): 275–316, 1997.
6. Max J Egenhofer. Toward the semantic geospatial web. In *10th ACM International Symposium on Advances in Geographic Information Systems*, pages 1–4, McLean, VA, USA, 2002.
7. Max J Egenhofer and John R Herring. Categorizing binary topological relations between regions, lines, and points in geographic databases. Technical Report 94-1, University of Maine, National Center for Geographic Information and Analysis, 1994.
8. Claudio Gutierrez, Carlos Hurtado, and Alejandro Vaisman. Temporal RDF. In *2nd European Semantic Web Conference*, pages 93–107, Heraklion, Crete, Greece, 2005.
9. Claudio Gutierrez, Carlos Hurtado, and Alejandro Vaisman. Introducing time into RDF. *IEEE Transactions on Knowledge and Data Engineering*, 19(2):207–218, February 2007.
10. Patrick Hayes. RDF semantics. http://www.w3.org/tr/rdf-mt/.
11. Hewlett-Packard Development Company. ARQ - a SPARQL processor for jena. http://jena.sourceforge.net/arq/.
12. Jerry Hobbs and Feng Pan. An ontology of time for the semantic web. *ACM Transactions on Asian Language Processing (TALIP): Special issue on Temporal Information Processing*, 3(1):66–85, 2004.
13. Krys Kochut and Maciej Janik. SPARQLeR: Extended SPARQL for semantic association discovery. In *4th European Semantic Web Conference*, pages 145–159, Innsbruck, Austria, 2007.
14. David Kolas, John Hebeler, and Mike Dean. Geospatial semantic web: Architecture of ontologies. In *1st International Conference on GeoSpatial Semantics*, pages 183–194, Mexico City, Mexico, 2005.
15. David Kolas and Troy Self. Spatially-augmented knowledgebase. In *6th International Semantic Web Conference*, pages 792–801, Busan, South Korea, 2007.
16. Manolis Koubarakis and Kostis Kyzirakos. Modeling and Querying Metadata in the Semantic Sensor Web: the model strdf and the query language stsparql. In Lora Aroyo, Grigoris Antoniou, Eero Hyvönen, Annette ten Teije, Heiner Stuckenschmidt, Liliana Cabral, and Tania Tudorache, editors, *Proceedings of the 7th Extended Semantic Web Conference (ESWC2010), Heraklion, Crete, Greece, May 30 - June 3, 2010, Proceedings, Part I*, volume 6088 of *Lecture Notes in Computer Science*. Springer, June 2010.
17. Joshua Lieberman. W3C geospatial incubator group. http://www.w3.org/2005/incubator/geo/.
18. Open Geospatial Consortium. Open geospatial consortium geospatial semantic web interoperability experiment. http://www.opengeospatial.org/projects/initiatives/gswie.
19. Jorge Perez, Marcelo Arenas, and Claudio Gutierrez. Semantics and complexity of SPARQL. In *5th International Semantic Web Conference*, pages 30–43, Athens, GA, USA, 2006.
20. Matthew Perry. Tontogen: A synthetic data set generator for semantic web applications. *AIS SIGSEMIS Bulletin*, 2(2):46–48, 2005.
21. Matthew Perry, Farshad Hakimpour, and Amit Sheth. Analyzing theme, space and time: An ontology-based approach. In *14th ACM International Symposium on Geographic Information Systems*, pages 147–154, Arlington, VA, USA, 2006.
22. Matthew Perry, Amit P. Sheth, Farshad Hakimpour, and Prateek Jain. Supporting complex thematic, spatial and temporal queries over semantic web data. In *2nd International Conference on Geospatial Semantics*, pages 228–246, Mexico City, Mexico, 2007.
23. Eric Prud'hommeaux and Andy Seaborne. SPARQL query language for RDF, W3C recommendation. http://www.w3.org/tr/rdf-sparql-query/.
24. Andrea Pugliese, Octavian Udrea, and V S Subrahmanian. Scaling RDF with time. In *17th International World Wide Web Conference*, pages 605–614, Beijing, China, 2008.
25. Wolf Siberski, Jeff Z. Pan, and Uwe Thaden. Querying the semantic web with preferences. In *5th International Semantic Web Conference*, pages 612–624, Athens, GA, USA, 2006.
26. Raj Singh, Andrew Turner, Mikel Maron, and Allan Doyle. GeoRSS: Geographically encoded objects for RSS feeds. http://georss.org/gml.

27. Philip D. Smart, Alia I. Abdelmonty, Baher A. El-Geresy, and Christopher B. Jones. A framework for combining rules and geo-ontologies. In *1st International Conference on Web Reasoning and Rule Systems*, pages 133–147, Innsbruck, Austria, 2007.
28. Yannis Theoharis, Vassilis Christophides, and Gregory Karvounarakis. Benchmarking database representations of RDF/S stores. In *5th International Semantic Web Conference*, pages 685–701, Galway, Ireland, 2005.

Chapter 4
Spatial Cyberinfrastructure: Building New Pathways for Geospatial Semantics on Existing Infrastructures

Francis Harvey and Robert G. Raskin

Abstract Spatial data infrastructures (SDI), with technological and conceptual roots stretching back multiple decades, are moving into a new era through the development of spatial cyberinfrastructures (spatial CI) that account for geospatial semantics. While the technology and concepts share many similarities, spatial cyberinfrastructures distinctly focus on the provision of information to support scientific knowledge sharing. These cyberinfrastructures are increasingly connected into an ecology of scientific knowledge sharing based on the formalization of geospatial semantics and support for shared knowledge and collective intelligence. We trace the development of cyberinfrastructures from spatial data infrastructures as the potential framework for geospatial semantical interoperability. The chapter also points to substantial semantic research challenges and the potential of spatial cyberinfrastructures.

4.1 Introduction

Spatial data infrastructures (SDI) provide a framework for organizing data and services, typically at the national level. These frameworks have been used to support application areas such as public administration [1–11] and more recently, the sciences [12, 13]. This chapter examines the distinct issues of spatial cyberinfrastructure as information communication technology for the organization of scientific knowledge and enabling geospatial semantics. The chapter engages the formalization of semantics through the creation of de facto and authoritative ontologies. The review considers both novel dimensions of cyberinfrastructure as well as established aspects of scientific knowledge sharing.

F. Harvey (✉)
Department of Geography, University of Minnesota, Minneapolis, MN, USA
e-mail: fharvey@umn.edu

N. Ashish and A.P. Sheth (eds.), *Geospatial Semantics and the Semantic Web:* *Foundations, Algorithms, and Applications*, Semantic Web and Beyond 12, DOI 10.1007/978-1-4419-9446-2_4, © Springer Science+Business Media, LLC 2011

The concept of cyberinfrastructure (CI), coined at the start of the twenty-first century to denote the development of capabilities to address state-of-the-art challenges of science and society, could have been defined at earlier stages of information technological development to express the potential of that particular era to articulate and implement spatial concepts. Different epochs have had different technologies for sharing knowledge. In the past 50 years we have gone from paper maps to web-mapping as the most common medium for sharing geographic information. Cyberinfrastructure takes off from technologies that are undergoing considerable transformation as computational abilities have increased many-fold. Technologies of the mid-twentieth century promised the abilities to carry out calculations not previously possible through large, centralized computers. Minicomputers of the 1970s enabled a modest research lab to own its own computer hardware and tailor its capabilities to the unique needs of the lab. By the 1980s, microcomputers enabled further miniaturization to the desktop-level and associated networks to readily share code and data. The 1990s saw the birth of the Internet and a new capability for any motivated person to publish content to the net-aware population at large. The initial decade of the twenty-first century brought these services to the masses, delivering social networking and new communication paradigms to a significant fraction of the global population. Each generation has seen a marked advance in computational capabilities and their contributions to challenges of science and society. In this sense, we see no future ossification of cyberinfrastructure, but an ongoing process characterized by specific local and global solutions interconnected in an ecology of scientific knowledge. Cyberinfrastructures exist in parallel with the establishing infrastructures of scientific knowledge including libraries, archives, journals, museums and other modes of storing, maintaining, and accessing scientific information.

Along this path, *spatial* CI has not only kept up with broad CI advancements, but often resided on its leading edge, from the Desktop GIS software of the 1980s [14–18], to on-line mapping applications of the 1990s [19–27], to the series of Where 2.0 Conferences that demonstrated the power of mashups, to the geotagging (georeferencing) of photos using Flikr, to the infusion of GoogleEarth technology for understanding of place names into the Google search engine. The spatial infrastructure is a key part of the 80% below the surface of Web 2.0 that contributes explicitly to scientific progress and enabling geospatial semantics. Thus, a consistent theme across the CI evolution has been new hardware or software yielding new service capabilities and the evolving institutional arrangements to exploit the new technology for science.

Each of these advancements brought with it an infusion of *shared knowledge* and *collective intelligence*. For research in the sciences, the motivation for collecting and analyzing spatial data is the contribution to scientific knowledge. This component of CI represents a third leg that complements and exploits the hardware and software advancements. As any formal definition of knowledge is nebulous, and efforts to provide a metrics for knowledge advancement at the societal level have proven to be challenging, institutional needs and priorities drive specific developments of the infrastructure. Figure 4.1 characterizes this evolution from the perspective of

	Data		Information			Knowledge	
Basic Elements	Bytes	Numbers		Models		Facts	
Storage	File	Database		Geodatabase		Ontology	Mind
Volume	High					Low	
Density	Low					High	
Services	Save	Discover	Visualize	Overlay	Infer	Understand	Predict

Syntax   *Semantics*

Fig. 4.1 Data to knowledge. As data is transformed to knowledge, it is summarized and contextualized, its volume is reduced, and its density (information or knowledge content per byte) increases

the transformation of data to information to knowledge. As the hardware/software components of CI advanced, a corresponding evolution on the data-to-knowledge continuum accrued; mid-twentieth century technology provided data, late 20th technology provided information, and the twenty-first century is seeing the emergence of knowledge. Spatial CI must resolve persistent technical and organizational issues to support long-term and multi-scalar science to achieve this promise. The fundamental opportunities of Web 2.0 enabling geospatial semantics affords new potential for developing paradigm-transcending ways of sharing knowledge.

As knowledge is inherently dynamic and expanding; spatial CI must consider a constantly changing field of knowledge. The knowledge component of CI is represented by the potential collective intelligence of the web, its blogs, tweets, and other social environments. Examples of spatial CI that exploit this knowledge can be found in areas such as geotagging, real-time traffic, and Google Street Map photos. We can think of this as a geospatial semantic web as a mature development of spatial CI, where geospatial literacy is greatly enhanced as a result of the collective sharing of geospatial experiences.

The knowledge capture repository is the *ontology*. An ontology provides shared understanding that is accessible to both humans and computers (in a form that is a compromise to both). The standardization of ontology languages by the W3C has enabled anyone with a web site to contribute to (and extract from) an emerging knowledge commons. Scientific communities will be important beneficiaries of shared knowledge infrastructures. The ontologies of the Semantic Web for Earth and Environmental Terminology (SWEET) are a typical example, in this case applicable to the relatively broad field of Earth system science. This ontology set includes concepts of science, data, and their applications, and it serves as a unifying framework for any resource that can be semantically associated. SWEET is designed to scale effectively as it grows by classifying knowledge by characteristic levels of abstraction, and by exploiting dependencies across concepts. SWEET also serves to break down any language barriers challenges of multi-disciplinary collaborations. As with related spatial cyberinfrastructure projects, we need to consider knowledge in these webs as the result of ongoing processes to learn about ontologies, then create them, and finally use them and integrate them into their institutions. This is also a process of recognizing and representing the knowledge of their respective communities, which transforms scientific practices and methods along with knowledge. Studies of the multi-faceted development of SDI emphasizing the cultural and political dimensions of large infrastructure projects hold important lessons for considering the development of scientific infrastructures [28–31].

4.2 Digital Earth, Virtual Globes, and Spatial Data Infrastructures

Taking a step back to the origins of spatial web concepts, the vision of a Digital Earth articulated by Former US Vice-President Al Gore in 1999, was prescient of this new paradigm, where large bodies of knowledge distributed across the network can be linked through geographic location. The Digital Earth foresaw a body of georeferenced information accessible to the fingertips, breaking down barriers to exchanges previously limited by geographic boundaries. In this vision, geography provides the unifying theme for broad access to georeferenced social, environmental, and physical information about the planet. This vision became lost in the US political events of the 2000–2009 era, which featured a contested election, an attack on American soil, and a prolonged war. Instead, the realization of the vision proceeded along two other fronts: (1) China took the lead through its *Beijing Declaration on Digital Earth* in 1999 which inspired several related activities: biannual International Symposiums on Digital Earth, the International Society for Digital Earth, and the International Journal of Digital Earth and (2) private US companies took advantage of new high-storage and high-bandwidth CI to deliver virtual globe products with massively large background data such as: GoogleEarth, VirtualGlobe. These tools place unprecedented capabilities into the hands of average world citizens and opportunities to browse and explore the planet in-depth from the comfort of one's home. These capabilities remain key to the experience and potential of newer applications such as GoogleOcean, GoogleMars, and GoogleMoon.

Virtual globes integrate heterogeneous geospatial information; but these accomplishments have also raised expectations for what such a system should accomplish [7]. Of particular interest are the web-based services that can be carried out, some of which will be discussed later. Although the terms Digital Earth and Virtual globes are nearly synonymous in most uses, distinctions in these meanings can be found in [11].

4.2.1 Spatial Data Infrastructures

The Digital Earth vision has features in common with the spatial data infrastructure (SDI) concept developed by various national government bodies to consolidate and promote geographic information sharing across agencies and stakeholders [32, 33]. An SDI is a technical and organizational arrangement that standardizes spatial referencing systems to ensure that multiple spatial/temporal coordinates, datum, and map projections can be reconciled. Examples include the US National Spatial Data Infrastructure (NSDI) and the Infrastructure for Spatial Information in the European Community (INSPIRE), However, no such infrastructure exists for the semantics of phenomenon, something referred to as a semantic reference system. Gore's vision did not explicitly include such a semantic component, so successes in this arena will surpass the initial expectations of the Digital Earth.

A key outgrowth of SDIs is the establishment of specifications and standards that support the integration of spatial data. As Keyhole Markup Language (KML), used for Google Earth applications, became widely utilized, the Open Geospatial Consortium (OGC) became custodian of the specification to ensure widespread participation and acceptance. The merging of KML with the Geographic Markup Language (GML) will provide further benefits to the geospatial community in disparate applications. These standards effectively enable spatial data to be stored and resources shared within this framework.

Advancements with virtual globes suggest that the corresponding spatial *knowledge* infrastructures will be global in scope rather than constrained as national or regional resources. Underlying each virtual globe application is a global, multi-resolution, tiled map terabytes in size, which enables exploration of a limited extent or resolution at any one time.

4.2.2   Spatial Concepts in RDF/OWL

The World Wide Web Consortium (W3C) has adopted several languages for ontology representation that are expressible in XML - the Resource Description Framework (RDF) and the Web Ontology Language (OWL). RDF and OWL allow some possibilities for representing geospatial semantics, although with some complexity. Each language is based on the paradigm of facts as triple relations in a subject-verb-object structure. The *class* is the abstract representation of a subject, and when instantiated, is an *individual*. The *property* is the representation of a verb, which provides the semantic relation between concepts. An object may be represented as a class, individual, or a numerical entity, depending on the property definition. The *subclass* and *subproperty* constructs inherit all attributes of their parent class and property, respectively, providing the key element of scalable dynamic growth.

Current RDF/OWL constructs are relatively primitive, and representation of geographic features and relations requires the use of some workarounds to address its deficiencies. For example, the requirement of triple relations is limiting, forcing any quadruple relation to be expressed as a pair of triples. The ambiguities of the *part-of* relation are well known (e.g., six possible interpretations are noted by Winston, et al., 1987, only one of which is of the geographic sense). Community standards help to homogenize the various possible methods of expression and add clarity, much as natural languages do on an everyday basis.

Geospatial semantics will ultimately enable new automated spatial reasoning services including personal agent-based services, real-time automated trip planning, and decision-making using sensor networks.

4.3   Mash-Ups, Dynamic Content and Automated Services Using Digital Globes: Challenges for the Semantically Enabled Geospatial Web

4.3.1   Virtual Communities and Online Collaboration

Authentication is a key to enabling online collaborations. We take for granted that participants at a face-to-face meeting are who they say they are. Does the web of knowledge need to be limited to trusted participants? Wikipedia is often hampered by anonymous contributors with agendas, as a result of which, barriers to new entries have risen, reducing the amount of new material made available.

4.3.2   Geospatial Semantic Web Services and Catalogs

The Semantic Web is a Web where browser, crawler, and other tools understand content and can exploit that information to the benefit of users. This vision, often called Web 3.0, requires the type of knowledge classification and registration described above. For geographic information, the issues are complex due to the often implicit way geographical designators are handled. Additionally, commonly used designators are notoriously vague and spatially imprecise. Chicago means different things for different communities. Understanding and developing formal representations of these differences and associating the user with the appropriate community are important research challenges.

Place-based ontologies have the potential to supplement SDIs with comprehensive common sense knowledge about locations. The resulting SDI will be a centralized knowledge repository to support web services for personal, community, commerce, and official government use.

This type of SDI can be developed by drawing on data and metadata created for clearinghouses developed following specifications for metadata in the United States, Europe and elsewhere. The semantics of place-based ontologies can be instantiated through a data mesh [34] to facilitate researchers' development of shared knowledge and collective intelligence to address scientific questions.

Current development activities in the US Federal Government known as the Geospatial Platform are creating a new architecture for US federal government data that defines standards that semantic computing can directly connect to. As described in the 2011 US President's budget, the portfolio management based approach will support geospatial one-stop and place-based initiatives including increased use of data.gov with emphasis on reuse of architectural standards and technology. The challenge of governance is no small issue, but the take-up of standards points to opportunities for scientists to develop and rely on web services to access more and more information through semantically enhanced web services.

The access to US federal government geospatial data through web services has an over 10-year history and is established as common geospatial web services. The web map service specification created first in 1999 stands out. Moving towards semantically enabled search and access capabilities, the Office of Management and Budget's technical architecture working group is developing geospatial segment guidance for all federal agencies and is hosting a GeoCloud initiative. The cloud computing stack enables Infrastructure as a Service (IaaS) and Platform as a Service (PaaS) solutions. PaaS includes approaches to create geographic information services that can be bound with semantic search and access services in applications and other services. These approaches reflect existing information science work and applications developed in the past decade. Work by Fonseca [36], Lutz [38, 39], Wolniewich and Yue [37] highlights the recent contributions to creating a semantically enabled spatial infrastructure, in this selection of the many recently published articles on the geospatial semantic web especially with application for collections of geological data held by a variety of governmental and research institutions [35–39].

4.3.3 Managing Spatial Cyberinfrastructures

The development of geographic knowledge webs coupled with potentials for social networking and virtual Earth requires substantial technological and organizational infrastructures. The existing spatial data infrastructures provide a foundation for spatial cyberinfrastructures. Known limitations and the new challenges for cyberinfrastructures are important research challenges.

4.3.3.1 Existing Spatial Data Infrastructures

A significant resource for spatial cyberinfrastructures are the many spatial data infrastructures (SDI) with holdings for localities, regions, states, the nation, and internationally. While most of this data has been organized in databases, many of which are remotely accessible, important steps in creating metadata standards and clearinghouse architectures have taken place. Through national, state, regional, and thematic gateways on the Internet, for example MetroGIS's DataFinder (http://www.metrogis.org/data/datafinder/index.shtml) these rich data resources are readily available for spatial CI.

4.3.3.2   Cyberinfrastructure Research Challenges

The creation and development of shared knowledge and collective knowledge make key contributions to humanity's knowledge. Underlying the challenges for the spatial CI if a persistent problem of all societies: forgetting. Past human activities have never had the opportunities, been able to work with the unparalleled scientific data resources collected today, nor store such amounts of data. The traditional scientific practice of project-orientated work that commences with data collection and concludes with publication has emphasized a comparatively short-term retention of data and information. The traditional publication forms have emphasized the reporting of results and findings. Once project results are published the data frequently reaches the end of its life expectancy and is readily forgotten as science progresses. Forgotten data is of little benefit for cyber infrastructures, and significant challenges are instilling a new ethos of data documentation and dissemination. National research agencies have made important inroads on this issue. Web 2.0 extensions to spatial CI being developed lead to changes in technologies and organizations to assure spatial data, information, and knowledge are accessible for knowledge sharing and the development of collective intelligence.

4.4   Conclusions

Spatial CI remains key to supporting geospatial semantics and the resources for a broad range of activities. Research in this domain faces significant challenges that can be helped by bringing the lessons of SDI research on board and developing more robust knowledge formalizations for future needs.

References

1. Bamberger, W.J., *Sharing geographic information among local government agencies in the San Diego region*, in *Sharing Geographic Information*, H. Onsrud and G. Rushton, Editors. 1995, Center for Urban Policy Research: New Brunswick, NJ. p. 119–137.
2. Burmanje, D. (2005) *Spatial Data Infrastructures and Land Administration in Europe*.
3. Crompvoets, J., *National Spatial Data Clearinghouses*. 2006, Wageningen, The Netherlands: Wageningen University.
4. de Man, W.H.E., *Understanding SDI: Complexity and institionalization*. International Journal of Geographical Information Science, 2006. **20**(3): p. 329–343.
5. Georgiadou, Y. and J. Stoter, *SDI for public governance–implications for evaluation research*, in *A Multi-View Framework to Assess Spatial Data Infrastructures*, J. Crompvoets, et al., Editors. 2009, Space for Geo-Information (RGI): Wageningen. p. 51–68.
6. Johnson, R., Z. Nedovic-Budic, and K. Covert, *Lessons from Practice. A Guidebook to Organizing and Sustaining Geodata Collaboratives*. 2001, GeoData Alliance: Reston, VA.
7. Masser, I. and M. Wegener, *Brave new GIS worlds*, in *GIS Diffusion: The Adoption and Use of Geographical Information Systems in Local Government in Europe*, H. Masser, H. Campbell, and M. Craglia, Editors. 1996, Taylor & Francis: London.

8. Nedovic-Budic, Z., J.K. Pinto, and L. Warnecke, *GIS database development and exchange: interaction mechanisms and motivations*. URISA Journal, 2004. **16**(1): p. 16–29.
9. Obermeyer, N., *Reducing inter-organizational conflict to facilitate sharing geographic information*, in *Sharing Geographic Information*, H. Onsrud and G. Rushton, Editors. 1995, Center for Urban Policy Research: New Brunswick, NJ. p. 138–148.
10. Rajabifard, A., et al., *The role of sub-national government and the private sector in future spatial data infrastructures*. International Journal of Geographical Information Science, 2006. **20**(7): p. 727–742.
11. Tulloch, D.L. and F. Harvey, *When Data Sharing Becomes Institutionalized: Best Practices in Local Government Geographic Information Relationships*. URISA Journal, 2008. **19**(2): p. 51–59.
12. Gahegan, M., et al., *Connecting GEON: Making sense of the myriad resources, researchers and concepts that comprise a geoscience cyberinfrastructure*. Computers and Geosciences, 2009. **35**(4): p. 836–854.
13. Ribes, D. and G.C. Bowker, *Between meaning and machine: Learning to represent the knowledge of communities* Information and Organization, 2009. **19**(4).
14. Burrough, P.A., *Principles of Geographical Information Systems for Land Resource Assessment*. 1987, Oxford: Oxford University Press.
15. Chrisman, N.R. *Challenges for Research in Geographic Information Systems*. in *International Geographic Information Systems Symposium*. 1987. Arlington, VA: NASA.
16. Dobson, J., *Automated Geography*. The Professional Geographer, 1983. **35**(2): p. 135–143.
17. Robinove, C., *Principles of Logic and the Use of Digital Geographic Information Systems*. 1986, U.S. Geological Survey: Washington D.C.
18. Tomlinson, R.F. *Geographic Information Systems - A new frontier*. in *International Symposium on Spatial Data Handling*. 1984. Zurich.
19. Abler, R.F., *Everything in its Place: GPS, GIS, and Geography in the 1990s*. Professional Geographer, 1993. **45**(2): p. 131–139.
20. Armstrong, M., *On automated geography!* The Professional Geographer, 1993. **45**(4): p. 440–442.
21. Bonham-Carter, G.F., *Geographic Information Systems for Geoscientists. Modelling with GIS*. Computer Methods in the Geosciences. Vol. 13. 1994, Tarrytown: Pergamon.
22. Craig, W.J., *Why we can't share data: Institutional Inertia*, in *Sharing Geographic Information*, H.J. Onsrud and G. Rushton, Editors. 1995, Center for Urban Policy Research: New Brunswick, NJ. p. 107–118.
23. Desham, P., M.P. Armstrong, and K.K. Kemp, *Collaborative Spatial Decision Making*. 1995, NCGIA: Santa Barbara, CA.
24. Elwood, S. and H. Leitner, *GIS and community-based planning: Exploring the diversity of neighborhood perspectives and needs*. Cartography and Geographic Information Systems, 1998. **25**(2): p. 77–88.
25. Goodchild, M. *Cartographic futures on a digital earth*. in *International Cartographic Congress*. 1999. Ottawa, Canada: ICA,.
26. Reeve, D. and J. Petch, *GIS, Organisations, and People. A Socio-technical Approach*. GIS for beginners, ed. D. Reeve and J. Petch. 1999, London: Taylor and Francis. 214.
27. Yapa, L., *Why GIS needs postmodern social theory, and vice versa*, in *Policy Issues in Modern Cartography*, D.F.R. Taylor, Editor. 1998, Elsevier Science: London. p. xx.
28. Georgiadou, Y., S.K. Puri, and S. Sahay, *Towards a potential research agenda to guide the implementation of Spatial Data Infrastructures—A case study from India*. International Journal of Geographical Information Science, 2005. **19**(10): p. 1113–1130.
29. Puri, S.K., *Technological Frames of Stakeholders Shaping the SDI Implementation: A Case Study from India* Information Technology for Development, 2006. **12**(4): p. 311–331.
30. Crompvoets, J., et al., eds. *A Multi-View Framework to Assess Spatial Data Infrastructures*. 2009, Space for Geo-Information (RGI): Wageningen.
31. Mukherjee, F. and R. Ghose, *Complexities in GIS construction and spatial knowledge production in Dane county, Wisconsin*. Cartography and Geographic Informaiton Science, 2009. **36**(4): p. 299–314.

32. Masser, I., *GIS Worlds: Creating Spatial Data Infrastructures.* 2005, Redlands, CA: ESRI Press.
33. Onsrud, H., *Research and theory in advancing spatial data infastructure concepts.* 2007, Redlands, CA: ESRI Press.
34. Parastatidis, S., E. Viegas, and T. Hey, A *"Smart" Cyberinfrastructure for Research.* Communications of the ACM, 2009. **52**(21): p. 33–37.
35. Brodaric, B., *Geo-pragmatics for the geospatial semantic web.* Transactions inf GIS, 2007. **11**(3): p. 453–477.
36. Fonseca, F. and A. Rogriquez, *From geo-pragmatics to derivation ontologies: New directions for the geospatial semantic web.* Transactions in GIS, 2007. **11**(3): p. 313–316.
37. Wolniewicz, P., *Easily-accessible digital palaeontological databases – a new perspective for the storage of palaeontological information.* Geologos, 2009. **15**(3–4): p. 181–188.
38. Lutz, M. and D. Kolas, *Rule-based discovery in spatial data infrastructure.* Transactions in GIS, 2007. **11**(3): p. 317–336.
39. Lutz, M., *Ontology-based descriptions for semantic discovery and composition of geoprocessing services.* Geoinformatica, 2007. **11**: p. 1–36.

Chapter 5
Ontology-Based Geospatial Approaches for Semantic Awareness in Earth Observation Systems

Kristin Stock, Gobe Hobona, Carlos Granell, and Mike Jackson

Abstract Current work towards making earth observation systems semantically aware attempts to improve user experience by allowing more flexibility in the way that users interact with earth observation systems. Such improvements may occur directly by making data discovery more semantically-flexible, and indirectly in providing intelligent functionality that removes some of the load from users in interpretation of data and processes.

Semantic awareness in earth observation systems may be addressed from four different angles: semantics and information modelling; semantic data management; semantic data discovery and semantic data processing. Each of these areas is the subject of ongoing and developing research in the broader geospatial community, has been applied in a number of different situations and systems, and presents particular challenges for earth observation systems.

The Global Earth Observation System of Systems (GEOSS) is a large, global, heterogeneous earth observation system and provides a case study of the use of different methods for achieving semantic awareness in each of these four areas. Furthermore, an example architecture for an earth observation system that involves multiple aligned ontologies illustrates the challenges posed by real world, heterogeneous systems.

In combination, the review of related work, applications and challenges in each of the four areas, together with the GEOSS case study and example architecture provide an indication of the state of the art in semantic research as it applies to earth observation system. Furthermore, this summary provides a hint towards the future for semantics in earth observation systems and the need for additional work in this area.

K. Stock (✉)
Centre for Geospatial Science, University of Nottingham, Nottingham, UK
e-mail: Kristin.Stock@nottingham.ac.uk

N. Ashish and A.P. Sheth (eds.), *Geospatial Semantics and the Semantic Web: Foundations, Algorithms, and Applications*, Semantic Web and Beyond 12, DOI 10.1007/978-1-4419-9446-2_5, © Springer Science+Business Media, LLC 2011

5.1   Introduction

Interoperability is defined as:

> The capability to communicate, execute programs, or transfer data among various functional
> units in a manner that requires the user to have little or no knowledge of the unique
> characteristics of those units.
>
> Term 01.01.47, ISO/IEC 2382–1:1993 Information Technology –
> Vocabulary – Part 1: Fundamental Terms

Interoperability occurs at a number of levels, including (but not limited to) physical hardware components (computers, networks, etc); encoding and syntax (the syntactic level); semantics and pragmatics [40].

Whilst all of these levels of interoperability are important and have received varying levels of attention, this chapter focuses on semantic interoperability. Semantics in this context refers to the meaning of components (data or software) used in distributed information systems, and semantic interoperability is the ability of multiple software components to cooperate although implemented with different languages or interfaces [43]. Specifically, this means that software components and the people using them are able to understand and interpret the meaning of the services and data exchanged in a distributed environment [22].

In order for the semantics of data and web services (referred to as resources in the remainder of this chapter) to be interpreted by distributed software components and users, descriptive representations of those semantics are created to support the resources. A common aim is for these representations to be machine-readable and interpretable, so that software components can dynamically analyse the representations to assist with discovery of resources to meet user needs, or to automatically determine whether two resources are semantically similar and may thus be integrated, to provide appropriate translation services or to combine resources in useful ways. Semantic representations are thus potentially a useful tool in supporting the development of distributed systems.

There are a several existing approaches to the representation of resources, providing varying degrees of semantic richness, structure and formalisation. A number of methods have been proposed that attempt to define a formal specification that is based on group consensus, involving agreement on shared concepts within a domain [52]. These approaches usually adopt a hierarchical structure as the central method of knowledge organisation, and may also include semantic relations between concepts. The most semantically rich of these are ontologies, but other variations that also represent semantics structurally with less formal rigour include thesauri, controlled vocabularies and data models [36]. Ontologies are currently usually represented with Description Logics. Description Logics take a particular approach to the representation of semantics that allow the modelling of concepts, roles and individuals and can be formally defined using logic-based semantics [18].

In contrast to highly structured and consensus approaches to the representation of semantics like Description Logic ontologies, a number of other approaches have been developed with various goals, including flexibility, dynamism, context-sensitivity and semantic expressiveness. These approaches include those based on logic (for example, [3]), language [38,50], formal algebraic specifications [33] and conceptual spaces [17,46].

Ontologies are currently the most popular method for representing semantics, because while they lack flexibility and context-sensitivity and are limited in their semantic expressiveness, they are relatively easy to understand, they conceptually extend notions of shared vocabularies that users are already familiar with, they allow limited reasoning and they are supported by a range of commercial and open source tools.

Ontologies can add some semantic awareness to Geographic Information Systems (GIS) [1]. By formalising specifications of concepts, applications are able to reference the same resources when they refer to the same concepts. This provides different GIS with a 'shared understanding' of the meaning of concepts. In this chapter, we refer to the ability of a system to apply the meaning of a concept in algorithms or processes as 'semantic awareness'.

The Earth is a complex system with multiple phenomena interacting both directly and indirectly. For example, our daily routines, such as recycling plastic, indirectly influence the amount of carbon dioxide in the atmosphere, whereas bushfires directly influence the amount of atmospheric carbon dioxide. Earth observation (EO) systems [12] monitor these phenomena through in situ and remote sensors including space-based sensors, leading to vast amounts of environmental data. Inevitably, data collected by EO systems can vary significantly due to the variety of sensors and information systems employed. This heterogeneity can be evident even for EO data representing the same phenomenon. Not only is heterogeneity in EO systems evident within aspects of data collection, it also affects data retrieval due to differences between a user's conceptualisation and that of the EO system. Moreover, with international initiatives to build interoperable EO systems there is an extra level of heterogeneity between different EO systems. There is a need therefore, to enable EO systems to share definitions of phenomena such that the same data can be interpreted as referring to the same concepts. Although current efforts at integrating EO systems consider calibration and validation of data from different sensors, the complex nature of Earth system processes and the vast amount of EO data generated daily calls for more automation of data analysis between different but interoperable EO systems.

This chapter focuses on the application of ontologies in establishing semantically-aware EO systems. Specifically, it aims to answer the question: how can capabilities demonstrated by ontology-based geospatial approaches enhance the semantic-awareness of EO systems? To this end, the chapter provides a review of the issues involved in making EO systems semantically-interoperable and aware, by providing background on the state of the art research in the area, and presenting some examples of geospatial applications that illustrate the approach and are relevant to the above themes. The review is organised into the following sections:

Section 5.2: Semantics and Information Modelling
Section 5.3: Semantic Data Management

Section 5.4: Semantic Data Discovery
Section 5.5: Semantic Data Processing

Following the review, Sect. 5.6 presents a case study of the Global Earth
Observation System of Systems (GEOSS)[1] and discusses and illustrates each of the
issues reviewed in Sects. 5.2–5.5 in the GEOSS context. GEOSS is an international
initiative to create a coherent EO system by enhancing interoperability between
thousands of in situ, airborne and space-based sensors [28]. GEOSS provides an
interesting case study because it is made up of several interoperating EO systems,
which presents particular challenges for semantic awareness and interoperability.
Finally, Sect. 5.7 presents a high-level architecture that may be used to achieve
semantically-aware EO systems.

5.2   Semantics and Information Modelling

Information modelling is an important aspect of the representation of data in an
EO system. Information modelling involves defining the set of concepts that are of
interest within a domain. This process is usually driven by a set of use cases or an
application area to identify which concepts are relevant and important.

In addition to the simple identification of relevant concepts, in a semantically
aware EO system, the semantics (or meaning) of those concepts must be agreed
on and defined. This is usually done by the creation of an ontology. In a large,
heterogeneous EO system, a number of ontologies may be involved and each
may have its own terminology, set of concepts and definition of those concepts.
In this case, multiple ontologies may be created to represent the worldview of
each information community, and methods for handling these multiple ontologies
must be included in the Data Discovery (see Sect. 5.4) and Data Processing (see
Sect. 5.5) components of the EO architecture to ensure that these heterogeneous
conceptualisations are accommodated.

5.2.1   Review of Related Work

A body of research has been pursued in recent years to develop suitable methods for
geospatial ontology engineering, including the initial elicitation of concepts through
group discussion and graphical tools, the iterative refinement of the ontology
through individual and group participation and the formal definition of the concepts
in an appropriate ontology language.

---

[1] http://www.earthobservations.org.

The most widely used language for representing Description Logic ontologies within web-based information systems is the Word Wide Web Consortium (W3C) standard Web Ontology Language (OWL) [11]. OWL is based on the Resource Description Framework (RDF) [31]. RDF provides a basic structure to describe any kind of concept and its associations and attributes in the form of a triple (subject–predicate–object). OWL extends RDF by defining specific semantic modelling constructs, including particular types of semantic relations between concepts, and by defining a set of standard semantic constructs that may be used, makes the comparison and evaluation of the semantics of concepts easier. These international standards (OWL and RDF) are widely used and particularly useful for large, heterogeneous EO systems, because they provide a common means of exchange of information at both the syntactic and semantic level. A number of software packages have been developed to support the creation of OWL and RDF ontologies, the most popular open source options being Protégé[2] and SWOOP.[3]

Other related work on semantic information modelling for EO systems has addressed the definition of a set of requirements or guidelines to ensure that geospatial ontologies are developed in a sound manner than can support their efficient management and eventual use in applications, including appropriate structuring and content [29].

Finally, the development of ontologies to support EO systems may require a collection of different ontologies with different purposes. Most important for EO systems are upper level ontologies that provide fundamental ontological concepts that are domain neutral (for example, DOLCE [5] and SUMO [42]); domain ontologies that describe concepts in the appropriate earth science domain and application ontologies that describe the way in which a domain ontology may be used to achieve a particular purpose [18].

5.2.2 Applications

Within the realm of environmental science, a number of domain ontologies have been created by different information communities. For example, OWL has been adopted in the development of the Semantic Web for Earth and Environmental Terminology (SWEET) [44]. SWEET offers detailed descriptions of environmental concepts, populated with properties that indicate different types of relationships between concepts. SWEET currently offers 4,100 concepts in 125 modular ontologies organized into subjects such as Geology, Space, Hydrosphere. Another example of an environmental science ontology is GEMET, which offers a multilingual taxonomy of about 6,500 concepts. The only relationship depicted within GEMET is the parent–child or IS–A relationship. OWL-encoded ontologies such as SWEET

---

[2]http://protege.stanford.edu/.
[3]http://code.google.com/p/swoop/.

and GEMET offer a URL for uniquely identifying a concept. Within a semantic web, this enables all systems supporting and sharing concepts of a particular ontology to reference the same URLs when referring to the same concepts. GEMET may also be accessed through a web service and an Application Programming Interface (API), both of which are useful for accessing the most current, maintained version of the ontology, rather than storing a local copy.

SWEET and GEMET are examples of generic domain ontologies that have been developed with a view to supporting a wide range of different projects and applications. An example of the application of a group of more specifically targeted ontologies can be seen in the COMPASS[4] project. This project used the DOLCE upper level ontology to create a domain ontology for marine instruments containing approximately 200 concepts and adopted an existing domain ontology of the scientific domain (Brodaric et al. 2008), then created a series of about 80 application ontologies to define scientific resources and web services to support the scientific knowledge infrastructure demonstrated by the project [49]. In this way, various ontologies are usually applied in combination to support an EO system, particular if it is large, distributed and heterogeneous.

5.2.3 Challenges for Geospatial Semantic Awareness in EO

One of the complex challenges for EO systems is to enable the resolution of unique identifiers from OWL-encoded ontologies to unique identifiers of sensors, observations and measurements. Another challenge for interoperability between EO systems is the sharing of ontologies between different EO systems. Even with mechanisms for sharing ontologies between EO systems, an additional challenge is to enable inference from those ontologies through determination of semantic similarity [45]. Some of these issues are discussed in more detail in the following sections.

5.3 Semantic Data Management

In an EO system, metadata is used to describe resources available to users. Metadata is basic information that describes the resources. In an EO system, metadata is usually stored in a standards compliant catalogue, also known as a registry.

Traditional registries store basic metadata describing simple aspects of the content (for example, name, resource provider, date of creation or publication),

---

[4]http://compass.edina.ac.uk/.

but for semantically-aware EO systems, semantic registries are required. Semantic registries extend the basic metadata content of a registry with richer semantic content.

5.3.1   Review of Related Work

The standard that is used almost exclusively for EO systems is the Open Geospatial Consortium Catalog Service Specification [41]. This standard includes a number of different protocol bindings, the most popular being Catalog Services for the Web (CSW), which binds the generic standard to the HTTP Protocol. The Catalog Service Specification describes in generic terms the kinds of content that may be requested and returned and the ways in which the registry can be queried. To specify additional details that are needed to use the protocol in a real world application, application profiles are used.

Two main application profiles have been defined, both of which make use of another standard to define the implementation details of CSW. The ebRIM Application Profile for CSW shows how the ebRIM information model can be used with CSW to create a registry implementation [39], and the ISO 19115/19119 Application Profile for CSW shows how the ISO metadata standard may be used with CSW [54]. However, a semantically-aware EO system requires some semantic content to be stored in the registry, and neither of these standards specifies how this might be done. A number of open-source software products are available that implement CSW ebRIM and ISO application profiles, including GeoNetwork[5] and deegree.[6]

Research focusing on spatial semantic registries has registered OWL-S ontologies (see Sect. 5.5) and related application ontologies in a registry using the OGC ebRIM Application Profile for CSW, and applied semantic searching capabilities to identify concepts that have a subsumption relationship with the queried concept to discover and orchestrate web services [57]. Another approach has taken an ISO standard for Feature Type Catalogues that represent limited semantic information, also in a registry complying with the ebRIM Application Profile for CSW [51].

Most recently, a third application profile for CSW has been developed that focuses specifically on the creation of a semantic registry. This OWL Application Profile for CSW defines a CSW compliant information model that reflects the OWL and RDF information models, and differs from previous approaches in that it accesses OWL content in its native OWL form, without duplicating or translating the content to suit some other information model [48, 49].

---

[5]http://geonetwork-opensource.org/.
[6]http://www.deegree.org/.

5.3.2   Applications

Numerous applications of CSW registries complying with either the ISO 19115 or the ebRIM application profile have been developed. However, applications that include semantic content in the registry are rare.

The most common approach to the combination of registries and semantic content has been to use ontologies to find concepts relating to a specified search concept outside the registry, and then search for services related to all of those concepts in an OGC-compliant registry [4, 30, 53]. In another approach, Hobona et al. [24] propose implementation of a mediator to match concepts from an ontology to keywords in metadata offered by CSW.

One exception to this is the COMPASS Project, which created and demonstrated a semantic registry complying with the OWL Application Profile for CSW introduced in Sect. 5.3.1 in the context of a geospatial knowledge infrastructure to support scientists working in the coastal and marine domain [49].

5.3.3   Challenges for Geospatial Semantic Awareness in EO

A particular challenge posed by large, heterogeneous EO systems is the need for multiple ontologies that may come from different sources. For example, an EO system that explores environmental impacts across multiple domains may need ontologies from different information communities in different specialist areas. In this case, the ontologies must be related to each other in some way (either by alignment or merging) in order to operate as a coherent whole.

5.4   Semantic Data Discovery

Ontologies can be used to find terms that are semantically related to a user's selected or entered term and thus ensure more effective discovery of semantically appropriate resources.

5.4.1   Review of Related Work

Portals used for discovery of geospatial resources are typically refered to as geoportals [37]. The search capability of geoportals parses and filters an index of metadata stored in a registry by location (for example, bounding coordinates), time period and thematic keywords. In semantically-aware EO systems, this index also includes semantic content (see Sect. 5.3).

The core of semantic discovery involves the identification of semantically similar concepts in an ontology or group of ontologies. Modelling the similarity of the meanings of concepts is a key feature of semantically aware EO systems. This capability is referred to as semantic similarity computation or semantic matching. There are several different approaches for estimating semantic similarity. The determination of semantic similarity is dependent on the form of the semantic description. Schwering [46] offers a classification of the different approaches into models based on networks, features, geometry, alignment and transformations.

5.4.2 Applications

Examples of the application of ontologies in data discovery include SPIRIT [16] and STORM [23]. SPIRIT adopted both a geographic ontology and a domain ontology, with the former offering 125,812 places and the latter offering 2,223 concepts used in the tourism sector. STORM adopted the Wordnet ontology, a linguistic ontology that offered over 50,000 words and 40,000 phrases, collected into over 70,000 sense meanings and developed by the Cognitive Science Laboratory at Princeton University [1,8]. Currently EO initiatives such as GEOSS have developed mechanisms for query expansion, the most adopted strategy to support semantic data discovery, which involves retrieving terms that are semantically-related to a query term and including the additional terms in a search filter [26].

5.4.3 Challenges for Geospatial Semantic Awareness in EO

Section 5.3.3 introduced the challenge of multiple ontologies within an EO system, and the need for special approaches to resource discovery in multiple aligned ontologies. In this case, search and discovery algorithms must be modified to not only identify semantically similar concepts within the same ontology, but also to identify semantically similar concepts across different, aligned ontologies. This is discussed in more detail in Sect. 5.7.

5.5 Semantic Data Processing

The ability to process and operate over geospatial data in analysis tasks is an intrinsic characteristic of GIS. Desktop GIS tools and geospatial web services provide a well-know set of geoprocessing functions that have normally been used to process local data sets. In general, web services technologies have facilitated data integration and promoted interoperability among heterogeneous distributed information sources using standards-based specifications.

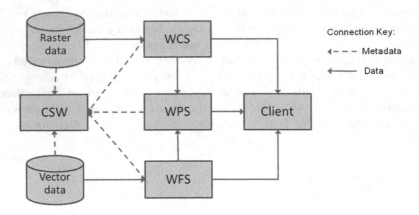

Fig. 5.1 Web services in a services oriented architecture

EO systems also follow the service-oriented architecture paradigm reflected in the widespread use of geospatial web services. These services allow users to access, manage, and process geospatial data in a distributed manner. The OGC is developing geospatial web services specifications by adapting or extending common-purpose web service standards, to facilitate data transformation, processing and integration, including for instance Web Feature Service (WFS) [55], Web Map Service (WMS) [10], Web Coverage Service (WCS) [56], Catalogue Services for the Web (CSW) [41], and Web Processing Service (WPS) [47]. Various open source software tools are available to assist users in creating resources based on these specifications, including GeoServer (which supports WFS, WMS and WCS)[7] and the 52°North[8] (which provides tools to support the creation of WPSs for different purposes).

In this sense, EO systems provide the infrastructure in which geospatial web services play a facilitating role to wrapping and abstracting data sources and integrating geospatial data and services. Figure 5.1 shows how the various OGC web services may be combined with geospatial data in Services Oriented Architectures.

5.5.1 Review of Related Work

One step towards the standardization of access and binding to service interfaces needed to offer generic geospatial processing services over the Internet is carried out by the OGC WPS specification, which provides service interfaces to deal with data processing aspects by either creating them from scratch or wrapping existing off-line functionalities as web services.

---

[7]http://geoserver.org/display/GEOS/Welcome.
[8]http://52north.org/.

The automation of processing data services is possible within the context of the Semantic Web applied to geospatial web services [7]. In this context, the combination of ontologies, relationships, and methods for ontology-based descriptions of geospatial data and services provides a formal framework for semantically-aware processing of geospatial data [34]. The predominant method to semantically describe geospatial web services is OWL-S. In short, OWL-S is an upper-ontology based on OWL that models the characteristics of web services, such as input and outputs parameters and functional aspects, which can be used to create semantically enriched web service descriptions and compositions.

Geoprocessing workflows are orchestrated sets of web services that perform particular geoprocessing functions (for example, coordinate transformations, map generalisation). Such orchestration of services is referred to as service chaining [2] and can enable the enactment of complex workflows [25] and allow geoprocessing functions created by remote providers to be used in multiple processes (thus avoiding reduced development effort).

5.5.2 Applications

Yue et al. [57] adopted OWL-S for supporting the automated construction of geoprocessing workflows. Their implementation extended the ebRIM profile of the OGC Catalogue Service for the Web (CSW) specification [41] to include OWL-S.

In contrast, Lemmens et al. [35] adopted WSDL-S for annotating semantically geoprocessing workflows. Semantic annotations enable linking non-semantic web services descriptions and specific vocabularies (e.g. concepts, relationships). WSDL-S, a common enabler for semantic annotation, is a variation of the Web Service Description Language (WSDL)[9] that includes references to OWL concepts. WSDL has been adopted by standardisation organisations such as OASIS, W3C and recently by the OGC. Lemmens et al. [35] observed that there are fewer enactors that support OWL-S than those supporting WSDL-S. They concluded that WSDL-S requires less effort to implement than OWL-S.

Fitzner et al. [15] propose a novel approach for annotating semantic geoprocessing services based on functional descriptions. The authors suggest the use of conjunctive queries in a logic programming language in order to formalise more precisely dependencies between the types of inputs and outputs, and the functionality itself of a WPS service. The authors conclude that the notion of conjunctive queries is quite new and not supported by current semantic web services approaches such as OWL-S.

---

[9]http://www.w3.org/TR/wsdl.

5.5.3 Challenges for Geospatial Semantic Awareness in EO

One challenge in the area of data processing posed by EO systems arises from the need to provide semantic descriptions of the existing multitude of geoprocessing tools. Once such semantic descriptions have been documented in OWL or in a similar ontology representation, it will then be necessary to provide a solution that offers improved usability and efficiency in the orchestration of geospatial web services for EO systems. Usability and efficiency are key issues because semantically-aware workflow enactors would need to compete with existing procedures adopted in EO systems.

Also, data provenance or lineage (ISO 19115) is recently gaining attention within the EO community [58]. As the number of available geoprocessing grows and geoprocessing workflows become widespread, it will be necessary to trace back the data sources and transformation processes used in generating EO products. Reliability and trust will be immediate issues that arise as a result of incorporating data provenance mechanisms into EO systems.

5.6 GEOSS Case Study

The GEOSS programme provides a useful case study of a large system of heterogeneous Earth observation systems, for which most of the issues reviewed in this paper thus far are pertinent, and to which most of the challenges raised apply. This section provides some background on the GEOSS project and the system, and then addresses each of the aspects of semantically-aware EO systems raised in Sects. 5.2–5.5 in the context of the project.

5.6.1 Overview

The Group on Earth Observations (GEO) coordinates the development of GEOSS, a programme to provide a conceptual, technical and operational framework for enabling multiple EO systems to work collaboratively. GEOSS is not intended to bring all of the world's EO systems into a single monolithic system, but to enhance interoperability between various EO systems such that they can offer comprehensive, coordinated, and sustained Earth observations. From a technical perspective, GEOSS consists of several contributed components such as web services and sensors.

In addition to the multitude of contributed components and services, GEOSS offers a shared infrastructure. The GEOSS Common Infrastructure (GCI) consists of the GEO Web Portal, community portals, clearinghouses for searching data, information and services, and registries containing information about GEOSS

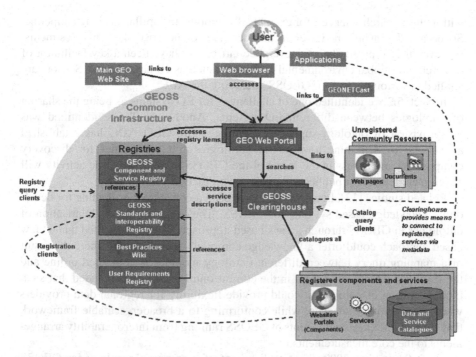

Fig. 5.2 GEOSS common infrastructure (Source: GEOSS AIP-3 Architecture)

components, standards and best practices. As illustrated in Fig. 5.2,[10] the GCI operates as a hub facilitating operations by users and applications alike. A key enabler of GCI operations is the suite of interoperability arrangements adopted in the development of the Architecture Implementation Pilot (AIP). Interoperability arrangements can be considered agreements between different parties on how their contributed services can operate cooperatively. They may include, for example, interoperability standards (such as those by the OGC) and business-to-business (B2B) rules that organisations may arrange bilaterally.

5.6.2 Semantics and Information Modelling in GEOSS

Sensors are the main devices for data collection within GEOSS. They can be in situ, airborne or space-based. In situ sensors are arguably the most widely used, for example, meteorological sensors detecting wind speed and direction are found in most countries. Borehole/well sensors monitoring groundwater quality are another example of in situ sensors. Airborne sensors have been adopted mainly

---

[10]http://www.earthobservations.org/documents/cfp/20100129_cfp_aip3_architecture.pdf.

within the research sciences, for example, for monitoring spillages along a pipeline. Space-based sensors are fewer in number due to the cost of such investments, however, they typically benefit several countries and have been a key facilitator of research into global environmental challenges such as climate change. Sensors are applied in various aspects of GEOSS, including the AIP.

In Sect. 5.2 we identified one of challenges for EO systems as being the sharing of ontologies between different EO systems. Another challenge identified was the provision of ontology-supported inference. The GEO ADC has established an activity to determine how to use ontologies and taxonomies for discovery composition and access in the frame of the GEOSS architecture. The activity will develop guidelines for publication on the Best Practice wiki.[11]

Related studies [13], however, have proposed a conceptual grounding for developing knowledge-based systems for GEOSS. They suggest the modularisation of ontologies in GEOSS through a view based approach. They highlighted that a view based approach could offer a knowledge compilation mechanism where the result of a mapping query between different ontology modules is computed offline and the result is added as an axiom in the current ontology. They concluded that such modularization of ontologies would provide flexibility for regional data providers when representing their data, while conforming to a broader, scalable framework that encapsulates key components of GEOSS ranging from interoperability arrangements to the core measurements.

As at September 2009, the establishment of a common ontology for GEOSS was in progress.[12] GEOSS however supports OGC sensor web standards, such as Sensor Model Language (SensorML)[13] and Observations and Measurements (O&M).[14] SensorML offers general models and XML encodings for sensors and observation processing. O&M offers general models and XML encodings for sensor observations and measurements. The standards are part of the OGC's Sensor Web Enablement (SWE), a suite of sensor oriented standards.

5.6.3 Data Discovery in GEOSS

The GEOSS 10-year implementation plan (Group on Earth Observations 2005) identifies ISO 19115 as the metadata standard to be adopted by GEOSS. The GCI offers a web service interface based on the OGC Catalogue service standard (CSW) for enabling applications to search through the registries, using the ISO19115

[11]http://www.earthobservations.org/documents/committees/adc/200909_11thADC/AR-09--01d_Ontology_Taxonomy_Melbourne.pdf.

[12]http://www.earthobservations.org/documents/committees/adc/200909_11thADC/AR-09--01a_20090915.pdf.

[13]http://www.opengeospatial.org/standards/sensorml.

[14]http://www.opengeospatial.org/standards/om.

Application Profile for CSW (Khalsa et al. 2009). ISO19115 identifies the following fields as core elements: *Dataset title; Spatial representation type; Dataset reference date; Reference system; Dataset responsible party; Lineage; Geographic location of the dataset; On-line resource; Dataset language; Metadata file identifier; Dataset character set; Metadata standard name; Dataset topic category; Metadata standard version; Spatial resolution of the dataset; Metadata language; Abstract describing the dataset; Metadata character set; Distribution format; Metadata point of contact; Additional extent information for the dataset (vertical and temporal)* and *Metadata date*.

The GEOSS architecture defines two types of portals, namely, a GEO Web Portal and GEOSS Community portals. The former is the main gateway to GEOSS. The latter is the series of portals developed by different GEO members and participating organisations for addressing the requirements of a specific community, such as disaster management or air quality monitoring. The ESRI geoportal, one of the portals offered by GEOSS, offers ontology-based query expansion [14]. The ESRI geoportal provides a web application based on the GEMET thesaurus that enables a user to find records containing keywords associated with a search term, in addition to records containing the actual keyword queried by the user.

5.6.4 Data Processing in GEOSS

The ability to process EO data is a key requirement for making such data usable by decision makers. GEOSS registered components offer services for transforming collected data into usable data. This may involve fundamental processing such as coordinate transformations or more advanced processing such as the modeling of air pollution.[15] The modular nature of service oriented architecture, including that of GEOSS [21], enables the creation of composite processes from orchestration of multiple services [19]. Web services for geospatial data processing within GEOSS are based on ISO 19119.[16] Two of the standards based on ISO19119 [27] and adopted by GEOSS include the OGC Web Feature Service (WFS) [32] and Web Coverage Service (WCS) standards. The AIP has demonstrated the orchestration of web services within workflow enactments using the Business Process Execution Language for Web Services (WS-BPEL, normally referred to as BPEL).[17]

The semantic information associated with these web services is seldom available in OWL. As GEOSS is intended to leverage existing systems while developing new components, much of the semantics describing attributes of web services within GEOSS has been documented in interoperability arrangements (including international standards, service level agreements and bilateral agreements between

---

[15]http://www.ogcnetwork.net/system/files/Final_20090817_AIP_AQ_ER_1.3.pdf.

[16]http://www.earthobservations.org/documents/cfp/20100129_cfp_aip3_architecture.pdf.

[17]http://www.oasis-open.org/committees/wsbpel/.

contributed components). Interoperability arrangements, such as standards, are meant to be human-readable. It can be foreseen, however, that future interoperability arrangements will include mediator applications that support data processing, dissemination and reception between multiple components.

5.7 A High Level Architecture for Semantically-Aware in Earth Observation Systems

Following a review of the relevant issues involved in semantically-aware EO systems and their illustration through the GEOSS case study, this section presents an architecture that may be used for EO systems.

The architecture proposes an approach to the handling of multiple ontologies through ontology alignment. This architecture is appropriate for large, heterogeneous EO systems that operate across multiple domains and information communities, each that have their own ontologies under separate management and control. In such cases, it is impractical and often counter-productive to create a new, single ontology to drive the EO system.

It is necessary for the ontologies to be semantically related, and this can be achieved through ontology alignment. Ontology alignment involves the definition of semantic relations between terms from different ontologies. This can be effectively achieved with SKOS, the Simple Knowledge Organisation System. SKOS is an RDF based model for defining semantic relations both within a single ontology (or thesauri), and between different, multiple ontologies. For mappings between ontologies, it defines five semantic relations: exactMatch, closeMatch, relatedMatch, broadMatch and narrowMatch.

The aligned ontology architecture involves a set of aligned ontologies that are used to semantically annotate the resources available within the EO system, and can thus be used to drive semantically-aware resource discovery. Resource annotation means that ontology terms that reflect the content of the resource are used to tag or annotate the resource, and can then be used in resource discovery. There are two possible approaches to ontology alignment within this architecture, and the choice of approach then affects semantic annotation and consequent resource discovery.

The first option for ontology alignment involves the selection of a core ontology to which all other ontologies are aligned. This core ontology must then be used to annotate resources available in the EO system. Users can then discover resources using any of the aligned ontologies, and upon selection of an ontology term of interest, the mappings are followed from that term back to semantically related core ontology terms and the resources that are annotated with those terms. This option has the advantage that each ontology must only be aligned to one other ontology and is thus practical. However, it has the significant disadvantage that users cannot annotate their resources with terms from the other ontologies, which restricts somewhat the benefit of aligning those ontologies. Ontologies are most likely to be aligned because the core ontology has insufficient terms to cover a specialised area, and thus it is also likely that it would be useful to have the terms from those ontologies available for resource annotation. The inability to use terms from other ontologies for resource annotation makes discovery similarly limited, because although users can select terms from other ontologies, this is not matched by a related precision in resource discovery.

The reason for this limitation is that in most cases it is not possible to derive links between non-core ontologies based on 1 to 1 mappings from each ontology back to the core ontology. Of the five SKOS semantic mapping relations, only 1 (exactMatch) is transitive, so inferences about the other types of spatial relations are not possible.

The second option overcomes these problems and allows semantic annotation using terms from any aligned ontology, and consequently more fine-grained resource discovery. However, the disadvantage of this approach is that all ontologies that are to be used in the EO system must be aligned to each other ontology, which is a large and time-consuming task. While some ontologies may be easily aligned (for example, they may take the form of a branch that can be grafted on to a leaf of another ontology), others have extensive semantic overlaps and alignment effort is significant.

Some work has been done on developing methods for automatic ontology alignment (for example, [9]), and these methods can assist in making this second option more practical. Also, this option works best if there are few ontologies, rather than a large number.

Discovery of resources also requires special treatment in an EO system that uses multiple aligned ontologies. Resource discovery in semantically-aware EO systems commonly considers the identification of semantically related resources. That is, the user may identify a concept of interest and any semantically aware architecture will use some method to identify semantically related concepts as well, and consequently will also allow the user to discover resources that are connected to those semantically related terms. In an architecture with multiple aligned ontologies, discovery algorithms must go further than the usual approaches to semantic similarity matching, and must also follow the semantic links that are defined between ontologies to identify semantically-related concepts from all ontologies. That is, if the user selects a concept from any given ontology, all semantically-related concepts from that ontology and from other aligned ontologies in the infrastructure must be identified and returned to the user.

Thus the ontology alignment architecture involves the following interrelated components in addition to the normal infrastructure for an EO system (see Fig. 5.3):

1. A component to either perform automatic ontology alignment or to support manual ontology alignment (or a combination of the two)
2. A component to allow semantic resource annotation
3. A component to manage the ontologies that are being used (most likely a semantic registry)
4. A semantic discovery component

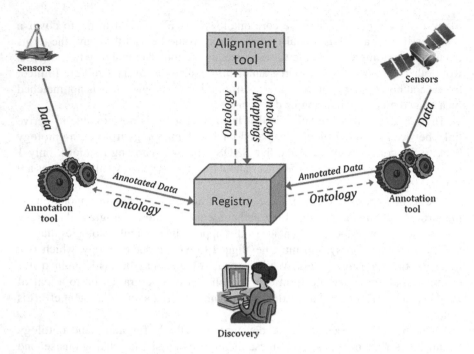

Fig. 5.3 Aligned ontology architecture

5.8 Conclusions

Large, federated, heterogeneous EO systems are becoming increasingly prevalent, due to developments in standards to support such infrastructures, increasing demand for integrated information to solve complex and wide-ranging environmental problems and increased economic pressures on system development requiring reuse of data and resources already developed by other agencies.

Interest in the semantic-awareness of EO systems has also increased in recent years due to recognition of the limitations of systems that do not support understanding of the meaning of earth observation objects and processes, the continuing development of research in the area and the gradual incorporation of semantic approaches that were originally confined to research arenas into real-world production systems such as GEOSS.

The chapter has provided a review of the state of the art in semantics research and applications as they apply to EO systems, as well as illustrating the use of the approaches in a globally significant system and providing a candidate architecture that is suitable in particular EO system contexts. Much of the work presented in this chapter applies to geospatial systems more generally, and some also to other types of information systems. However, EO systems are particularly characterised by their heterogeneity, semantic-diversity and geographic dispersion, as well as the presence of specialised data capture techniques such as sensors and a well-supported programme of standards development that makes the development of such systems possible.

The work described and presented in this chapter provides a snapshot in an ongoing process of development, and it is anticipated that developments will continue to gather momentum. Future developments may see the increased embedding of the use of ontologies in production EO systems, moving even further out of the research realm and into mainstream adoption. While ontologies are becoming increasingly popular and may be considered the current state of the art in semantic-enablement of EO systems, they have some deficiencies. Most importantly, it is not a trivial task to create an ontology for a particular domain (it takes significant time, domain expertise and ontology engineering expertise, usually involving a team of individuals), and even then, it only represents one view of the domain: that of a particular information community. Furthermore, the ontology representation is highly structured and precise, often requiring decisions to be made about categorisation and relations that do not reflect the ways in which people think about geographic concepts. Thus while ontologies represent a useful first step towards semantic-enablement and are helping to make users more semantically literate, they are only a first step in solving the problem of semantic representation.

Looking forward, while ontology work (both in the research and production environments) is likely to continue, it is expected that developments in non-ontology approaches to semantic representation may become more sophisticated and eventually augment the ontology work that is becoming established. These non-ontology approaches may do this by providing more flexibility and incorporating solutions to some of the more complex problems of geographic information that are still largely in the research realm, including dynamism, vagueness, context sensitivity, multi-lingual and natural language representations. Progress in combating existing challenges in recent years has been significant, and even more interesting challenges lie ahead!

References

1. Agarwal P (2005) Ontological considerations in GIScience. International Journal of Geographical Information Science 19: 501–536
2. Alameh N (2003) Chaining Geographic Information Web Services. IEEE Internet Computing, 7: 22–29
3. Bennett B, Mallenby D, Third A (2008) An Ontology for Grounding Vague Geographic Terms. In: C. Eschenbach, M. Gruninger, (eds.) Formal Ontology in Information Systems, Proceedings of the Fifth International Conference (FOIS08), IOS Press.
4. Bernard L, Einspanier U, Haubrock S, Hübner S, Klien E, Kuhn W, Less-ing R, Lutz M, Visser U (2004) Ontology-Based Discovery and Re-trieval of Geographic Information in Spatial Data Infrastructures. In Geotechnologien Science Report No 4.
5. Borgo S, Carrara M, Garbacz P, Vermaas PE (2009) A Formal Ontological Perspective on the Behaviors and Functions of Technical Artifacts. Artificial Intelligence for Engineering Design, Analysis and Manufacturing (AIEDAM), 23: 3–21

6. Brodaric B, Reitsma F, Qiang Y (2008) SKIing with DOLCE: toward an e-Science Knowledge Infrastructure. In: C. Eschenbach, M. Gruninger (eds.), Formal Ontology in Information Systems, Proceedings of the Fifth International Conference (FOIS08), IOS Press, pp.208–219.

7. Cardoso J, Sheth A (2005) Introduction to semantic web services and web process composition. In: Cardoso, J., Sheth, A. (eds.), Proceedings First International Workshop on Semantic Web Services and Web Process Composition (SWSWPC 2004). Springer, Berlin, Germany, pp. 1–13. LNCS 3387.

8. Chen C (2004) Information Visualization: Beyond the Horizon. Springer-Verlag, London.

9. Cruz IF, Sunna W, Makar N, Bathala S (2007) A visual tool for ontology alignment to enable geospatial interoperability. Web Semantics: Science, Services and Agents on the World Wide Web 5: 39–49

10. de la Beaujardiere J (2006) Web Map Service Implementation Specification. Open Geospatial Consortium Specification 06–042.

11. Dean M, Schreiber G, Bechhofer S, van Harmelen F, Hendler J, Horrocks I, McGuinness D, Patel-Schneider P, Stein L (2004) OWL Web Ontology Language Reference. W3C Recommendation 10 February 2004. http://www.w3.org/TR/owl-ref/. Accessed 12 June 2009

12. Durbha SS, King RL (2005) Semantics-Enabled Framework for Knowledge Discovery From Earth Observation Data Archives. IEEE Transactions on Geoscience and Remote Sensing, 43: 2562–2572

13. Durbha SS, King RL, Younan NH (2008) An Information Semantics Ap-proach for Knowledge Management and Interoperability for the Global Earth Observation System of Systems. IEEE Systems Journal, 2: 358–365

14. ESRI (2010) ArcGIS Server Geoportal Extension. http://geoss.esri.com/geoportal/webhelp/en/geoportal_931/index.htm#srch_ont.htm. Accessed 20 March 2010.

15. Fitzner D, Hoffmann J, Klien E (2009) Functional description of geoprocessing services as conjunctive datalog queries. Geoinformatica 15, (1), pp 191–221

16. Fu G, Jones CB, Abdelmoty AI (2005) Ontology-Based Spatial Query Expansion in Information Retrieval. Lecture Notes In Computer Science, 3761: 1466–1482

17. Gardenfors, P (2004) Conceptual spaces as a framework for knowledge representation. Mind and Matter 2: 9–27

18. Gomez-Perez A, Fernandez-Lopez M, Corcho O (2004) Ontological Engineering. Springer, London

19. Granell C, Díaz L, Gould M (2010) Service-oriented applications for environmental models: reusable geospatial services. Environmental Modelling and Software, 25: 182–198

20. Group on Earth Observations (2005) Global Earth Observation System of Systems GEOSS - 10-Year Implementation Plan Reference Document. ESA Publications Division, Noordwijk, The Netherlands

21. Group on Earth Observations (2010) GEOSS AIP Architecture. GEOSS Architecture Implementation Pilot (AIP)

22. Heiler S (1995) Semantic Interoperability. ACM Computing Surveys, 27: 271–273

23. Hobona G, James P, Fairbairn D (2006) Multidimensional Visualisation of Degrees of Relevance of Geographic Data. International Journal of Geographic Information Science, 20: 469–490

24. Hobona G, Fairbairn D, James P (2007) Semantically-assisted geospatial workflow design. In H. Samet, C. Shahabi & M. Schneider (eds.), Proceedings of the ACM international symposium on Advances in geographic information systems (ACMGIS) (pp. 194–201). ACM, New York

25. Hobona G, Fairbairn D, Hiden H, James P (2010) Orchestration of Grid-enabled Geospatial Web Services in Geoscientific Workflows. IEEE Transactions on Automation Science and Engineering, 7, (2), pp. 407–411.

26. Hogeweg M (2008) Portal Semantic Searches. In GEOSS AIP mailing list post of 30/11/2008 [AIP_PortalWG]

27. International Organisation for Standardisation (2005) ISO 19119:2005 Geographic information – Services.

28. Khalsa SJ, Nativi S, Geller G (2009) The GEOSS interoperability process pilot project (IP3). IEEE Transactions on Geoscience and Remote Sensing 47: 80–91
29. Klien E, Probst F (2005) Requirements for Geospatial Ontology Engineering. In: Toppen, F. and Painho, M. (eds.), Proceedings of the 8th Conference on Geographic Information Science (AGILE 2005), Estoril, Portugal, pp. 251–260
30. Klien E, Lutz M, Kuhn W (2006) Ontology-Based Discovery of Geo-graphic Information Services - An Application in Disaster Manage-ment. Computers, Environment and Urban Systems 30: 102–123
31. Klyne G, Carroll JJ (2004) Resource Description Framework (RDF): Con-cepts and Ab-stract Syntax. W3C Recommendation 10 February 2004. http://www.w3.org/TR/rdf-concepts/. Accessed 22 January 2009.
32. Kottman C (1999) The OpenGIS Abstract Specification Topic 14: Semantics and Infor-mation Communities. Version 4. OpenGIS Consortium Document 99–114. http://www. opengeospatial.org/standards/as. Accessed 20 March 2010
33. Kuhn W (1997) Approaching the Issue of Information Loss in Geographic Data Transfers. Geographical Systems 4: 261–276
34. Lemmens R (2008) Using Formal Semantics for Service within the Spatial Information Infrastructure. In: van Oosterom, P. & Zlatanova, Z. (eds.), Creating Spatial Information Infrastructures. Towards the Spatial Semantic Web, pp. 103–118. CRC Press.
35. Lemmens R, Wytzisk A, By R, Granell C, Gould M, van Oosterom P (2006) Integrating Semantic and Syntactic Descriptions to Chain Geographic Services. IEEE Internet Computing, 10: 42–52
36. McGuinness DL (2003) Ontologies Come of Age. In Dieter Fensel, Jim Hendler, Henry Lieberman, and Wolfgang Wahlster (eds.) Spinning the Semantic Web: Bringing the World Wide Web to Its Full Potential. MIT Press
37. Maguire DJ, Longley P (2005) The emergence of geoportals and their role in spatial data infrastructures. Computers, Environment and Urban Systems 29: 3–14
38. Mark DM, Egenhofer MJ, (1994) Modeling Spatial Relations Between Lines and Regions: Combining Formal Mathematical Models and Hu-man Subjects Testing. Cartography and Geographic Information Systems 21:195–212
39. Martell R (2005) OGC Catalogue Services – ebRIM (ISO/TS 15000–3) profile of CSW. Open Geospatial Consortium Specification 05–025r3
40. Morris C (1971) Writings on the General Theory of Signs. Mouton, The Hague
41. Nebert D, Whiteside A, Vretanos P (2007) OGC Catalogue Service Specification. Open Geospatial Consortium Specification 07–006r1
42. Niles I, Pease A (2001) Towards a Standard Upper Ontology. In Chris Welty and Barry Smith (eds.), Proceedings of the 2nd International Conference on Formal Ontology in Information Systems (FOIS-2001), Ogunquit, Maine, October 17–19, 2001.
43. Ram S, Park J, Lee D (1999) Digital Libraries for the Next Millennium: Challenges and Research Directions. Information Systems Frontiers 1: 75–94
44. Raskin RG, Pan MJ (2005) Knowledge representation in the semantic web for Earth and environmental terminology (SWEET). Computers & Geosciences 31: 1119–1125
45. Schwering A (2007) Evaluation of a semantic similarity measure for natural language spatial relations. In Proceedings of the International Conference on Spatial Information Theory (COSIT 2007), Melbourne, Australia.
46. Schwering A (2008) Approaches to Semantic Similarity Measurement for Geo-Spatial Data: A Survey. Transactions in GIS 12: 5–29
47. Shut P, Whiteside A (2005) Web Processing Service Implementation Specification. Open Geospatial Consortium Specification 05–007r4
48. Stock, Kristin (2009) OWL Application Profile for CSW 2.0. Open Geospatial Consortium Application Profile 09–010
49. Stock K, Robertson A, Bishr M, Stojanovic T, Ortmann J, Reitsma F, Medyckyj-Scott D (2009) eScience for Sea Science: A Semantic Knowledge Infrastructure for Marine Scientists. Proceedings of the 5th IEEE International Conference on e-Science, Oxford, UK, December 9–11.

50. Stock, Kristin (2010) Describing Spatial Relations Using Informal Semantics. Proceedings of GIS Research UK (GISRUK 2010), 14–16 April, London, UK.
51. Stock K, Atkinson R, Higgins C, Small M, Woolf A, Millard K, Arctur D (2010) A semantic registry using a Feature Type Catalogue instead of ontologies to support spatial data infrastructures. International Journal of Geographical Information Science 24: 231–252
52. Studer R, Benjamins VR, Fensel D (1998) Knowledge Engineering: Principles and Methods. IEEE Transactions on Data and knowledge Engineering 25:161–197
53. Vogele T, Spittel R (2004) Enhancing Spatial Data Infrastructures with Semantic Web Technologies. In Toppen, F. and Painho, M. (eds.) Proceedings of the 7th Conference on Geographic Information Science (AGILE 2004), 29 April – 1 May, Heraklion, Greece
54. Voges U, Senkler K (2005) Open GIS Catalogue Services Specification 2.0 – ISO19115/ISO19119 Application Profile for CSW 2.0. Reference number OGC 04–038r2
55. Vretanos P (2005) Web Feature Service Implementation Specification. Open Geospatial Consortium Specification 04–094
56. Whiteside A, Evans JD (2008) Web Coverage Service (WCS) Implementation Standard. Open Geospatial Consortium Specification 07–065r5
57. Yue P, Di L, Yang W, Yu G, Zhao P (2007) Semantics-based Automatic Composition of Geospatial Web Service Chains. Computers & Geosciences, 33: 649–665
58. Yue P, Gong J, Di L (2010) Augmenting geospatial data provenance through metadata tracking in Geospatial service chaining. Computer & Geosciences, 36: 270–281

Chapter 6
Location-Based Access Control Using Semantic Web Technologies

Rigel Gjomemo and Isabel F. Cruz

Abstract Location-based applications are an important case within context-aware applications. They pose interesting challenges when access control is considered for they must satisfy requirements arising from the mobility of both users and resources. Further challenges arise in collaborative environments where resources are shared by users of different organizations. In this paper we propose an access control framework based on the Role Based Access Control (RBAC) model where users and resources are abstracted as sets of attributes that include their geospatial position. In our framework, collaboration is achieved through the interoperation of the access control systems of the collaborating organizations. We use Semantic Web languages, namely OWL and SPARQL. We argue that their expressive power can model a wide range of RBAC policies. In particular, reasoning as provided by OWL supports both a standard enforcement mechanism and interoperation. We have implemented our framework and studied time performance as a function of the number of users and of the roles they can assume. Our implementation also features an interface that visually depicts users and resources on a map. As users move around, the set of actions that they can execute on the resources is shown.

6.1 Introduction

The growing computing power of mobile devices and the development of wireless communication infrastructures offer many opportunities to potentially access data from any location. This enhanced accessibility raises concerns about data security. Different techniques exist for ensuring data security, such as authentication, encryption, and access control. In this chapter we focus on access control and in particular on location-based access control.

R. Gjomemo (✉)
ADVIS Lab, Department of Computer Science, University of Illinois at Chicago,
Chicago, IL, USA
e-mail: rgjomemo@cs.uic.edu; rigelgjomemo@yahoo.com

N. Ashish and A.P. Sheth (eds.), *Geospatial Semantics and the Semantic Web:*
Foundations, Algorithms, and Applications, Semantic Web and Beyond 12,
DOI 10.1007/978-1-4419-9446-2_6, © Springer Science+Business Media, LLC 2011

Access control systems can be conceptually divided in three abstractions: access control policies, enforcement mechanisms, and access control models [22]. *Access control policies* are high-level specifications and requirements about which users are allowed to access which data and how that access can be performed. *Enforcement mechanisms* deal with the execution of access control policies when access to data is requested by users. Enforcement mechanisms include, for instance, authentication points (such as a login window) and lists containing the allowed actions (e.g., read, write for every file in a disk). *Access control models* are formal representations of access control policies and serve as an intermediate link between policies and enforcement mechanisms.

An example of access control model is the Role Based Access Control (RBAC) model, which represents access control policies in terms of users, resources, permissions, and roles [11, 18, 19]. A *permission* is a general concept that represents an action or a mode of access on a resource. A *role* is a concept that represents a job function and a set of permissions needed to perform that function. The concept of role is central in many systems that implement RBAC. Roles create an abstraction layer by decoupling users from their permissions thus facilitating management of access control in large organizations.

In location-based access control, which is a special case within context-aware access control, the geospatial positions of users and resources play a central role in deciding what resources users can access. Furthermore, data may not be the only resources available to mobile users. Depending on the application, other resources can be identified. A location-based access control system must address several challenges:

Multiple types of resources and actions Compared with other more traditional applications, resources (and actions allowed on them) may be rather heterogeneous. Consider, for example, a traditional access control system for files versus an access control system for health care emergency. In the former system, there are few resource types (e.g., files and directories) and few associated actions (e.g., read, write), whereas in the latter system there is a large variety of resources (roads, hospital rooms, ambulances) with a large variety of possible actions on those resources. A location-based access control system must be able to represent and enforce access control policies for a large variety of resources and actions.

User and resource mobility Users and resources are mobile and their location may change rapidly. Access to mobile resources by mobile users may depend on both user and resource locations. For example, a medical doctor at a particular location, such as the hospital where she works, may have permissions that are not recognized at another location, such as the hospital where she is visiting a family member. That is, in the former case her role is medical doctor and in the latter case it is visitor. Access may also depend on other dynamic conditions, such as usage and availability of resources. A location-based access control system must be able to incorporate and represent other contextual information about the environment, which may be relevant to access control.

Multiple ownership of resources Resources may belong to different organizations with autonomous access control systems and different user groups. In environments where organizations need to collaborate, such as large sports events or health emergency situations, organizations may need to share resources and users temporarily. A location-based access control system for collaborative environments must therefore be able to support integration of access control policies of different organizations and their enforcement mechanisms.

RBAC addresses the challenge of representing access control policies for multiple types of resources and actions via the concept of permission, which is a general application-independent concept. To address user mobility, several research proposals extend RBAC with the concept of location [1, 2, 5, 10, 17]. In these proposals, RBAC roles and permissions are associated with location constraints. Users and permissions can be assigned to roles only if users are at specific locations. Other research proposals address the problem of multiple ownership of resources and interoperation between RBAC policies [14, 20]. In these proposals, RBAC policies of different organizations are integrated into a global RBAC policy that regulates access to resources across organizational boundaries. However, these proposals deal with collaboration in the context of pure RBAC and do not consider later extensions to RBAC.

The research proposals mentioned so far only consider access control models. They deal with extension and integration of access control policies modeled with RBAC, without considering enforcement mechanisms challenges. In particular, when location is incorporated into RBAC and when mobile users and resources are considered, the conditions under which access is requested may be very diverse. Consequently, the enforcement mechanisms must be able to represent and enforce complex and diverse location-dependent conditions. This challenge may become more complex during collaboration since the enforcement mechanisms of collaborating organizations should be able to interoperate.

In this paper, we propose a framework for context-aware access control (and therefore for location-based access control) in collaborative environments. Our framework defines both an access control model and its enforcement mechanisms. Our access control model extends previous work on location-based RBAC by abstracting users and resources as sets of attributes and values. The attributes contain location of users and resources. We implement our system by using Semantic Web standard languages, in particular OWL [4] and SPARQL [16], which provide a high expressive power that is suitable to represent a wide range of RBAC policies. To support collaboration, we use the reasoning features of OWL as a standard enforcement mechanism for RBAC policies belonging to different organizations.

Our contributions are as follows:

Context-aware RBAC model We define an RBAC model extended with context information (e.g., location). We associate users and resources with contextual attributes that represent not only their location but also other aspects relevant to access control. We associate context constraints with roles and permissions. *Context constraints* specify conditions under which users and permissions may be assigned to roles. In particular, as users and resources move around and as their contextual attribute values change, the roles of those users also change and the permissions related to those resources are enabled or disabled.

Semantic Web implementation We define a way of representing RBAC using OWL and SPARQL, which uses reasoning for the enforcement of access control policies. In particular, we show how RBAC entities and assignment relations between them can be represented so that a reasoner can enforce those policies. Our implementation approach has both differences and similarities with other research proposals that use OWL to represent RBAC [6,8,12,13,21,23], which depend on the design choices made for the representation of RBAC entities with OWL constructs. in common regard design choices about which OWL constructs to use for representing which RBAC concepts. reasoning features to enforce RBAC policies, while some of these approaches do not. In addition, we define a procedure to help security administrators derive semi-automatically OWL representations of RBAC policies.

Interoperation framework Supported by the RBAC model that we define and by its OWL implementation, we define a new interoperation framework for access control systems belonging to collaborating organizations. In particular, we define OWL mappings between RBAC entities belonging to several organizations. The enforcement mechanisms of the interoperation framework are implemented by the same standard reasoning features that are used for the enforcement mechanisms of a single organization.

The rest of this chapter is organized as follows. In Sect. 6.2, we briefly describe the RBAC model and the Semantic Web languages used in our implementation. In Sect. 6.3, we describe our own approach: we describe a scenario and examples of access control policies and the extensions to RBAC that allow for the incorporation of location and context into our access control model. Finally, in this section, we describe our OWL representation of the extended RBAC model and the use of the reasoner as an enforcement mechanism. In Sect. 6.4, we describe the interoperation framework. In Sect. 6.5, we describe our implementation system. Section 6.6 discusses work related to both access control models and enforcement mechanisms. Section 6.7 contains conclusions and future work.

6.2 Preliminaries

We start by succinctly presenting the RBAC model and Semantic Web languages used in our framework.

6.2.1 RBAC Model

The RBAC model was originally introduced with the aim of easing management of access control policies for large organizations. The access control policies are specified in RBAC by using the following entities: *users, roles, resources, actions,*

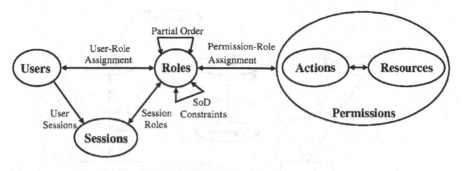

Fig. 6.1 Schematic depiction of the RBAC model

permissions, and *constraints*. A schematic representation of RBAC is shown in Fig. 6.1 [18]. In addition to the entities previously mentioned, it also shows how they are interrelated. In particular, the figure shows clearly the decoupling of users and of permissions via *roles*: users are assigned to roles, which in turn are associated with permissions. When users enter the system, they are assigned to *sessions* and assume the roles indirectly via the sessions. A partial order can be defined between roles, inducing a hierarchical structure that denotes inheritance of permissions (for example, a manager inherits the permissions of employee). *Separation of duty (SoD)* constraints state that two or more roles cannot be assigned to the same user at the same time. Other constraints (not depicted in the figure) may limit the number of users assigned to a role or the type of permissions associated with a role. Constraints may also be defined to specify the assignment of users to roles and of permissions to roles.

The formal definitions of the assignment relations between the RBAC entities are listed next [18], where the roles are denoted by R, the permissions by P, the users by U, and the sessions by S.

- $PA \subseteq P \times R$, a many-to-many permission-to-role assignment relation.
- $UA \subseteq U \times R$, a many-to-many user-to-role assignment relation.
- $user : S \rightarrow U$, a function mapping each session to a unique user.
- $RH \subseteq R \times R$, a partial order on R, called *role hierarchy*, denoted by $\geq$.
- $roles : S \rightarrow 2^R$, a function mapping each session to a set of roles.

The RBAC model is general and can be adapted to any domain by using roles and permissions specific to the domain.

6.2.2   Semantic Web Languages

In our framework we use the OWL [4] and SPARQL [16] languages. Knowledge about a domain is represented in OWL by using *classes*, *individuals* belonging to

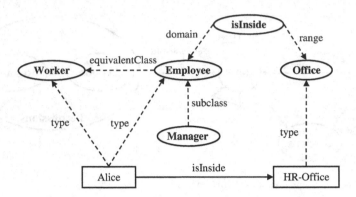

Fig. 6.2 An ontology with classes and properties (*oval nodes*), and individuals (*rectangular nodes*). OWL constructs are represented by *dashed lines* and user-defined entities are represented by *solid lines*

classes, and *properties*. OWL properties are either object properties or datatype properties. *Object properties* represent relationships between individuals, while *datatype properties* represent attribute values of individuals. OWL contains several constructs for composing complex classes from simple classes and describing complex knowledge about individuals and properties.

A set of OWL expressions describing entities of a domain is called an *ontology*. A simple example of an ontology about people in an office is depicted in Fig. 6.2. *Employee, Manager, Worker*, and *Office* are OWL classes and *isInside* is an object property that represents a relationship between *Employees* and *Offices*.

OWL reasoners operate on ontologies by inferring new relationships from existing ones. The basic reasoning tasks that are supported in OWL include: *classification*, which places an OWL class in the "proper place" in a class hierarchy, and *realization*, which finds the most specific class to which an OWL individual belongs [3]. To keep the algorithms that implement reasoning decidable, certain descriptive capabilities are excluded from OWL. For instance, it is not possible in OWL to express the fact that OWL individuals have the same value for a property (equivalent to a join on the property values of individuals), or to compose two or more properties into a single property. Consequently, OWL must be combined with other languages if the need for these descriptive capabilities arises. In our framework, we combine OWL with SPARQL [16].

SPARQL is an SQL-like query language for ontologies. SPARQL queries are composed of clauses similar to SQL clauses, such as *SELECT* and *WHERE*. The *WHERE* clause of SPARQL contains a set of *triples* that describes a portion of the graph of the ontology. Triples can contain variables that are bound to entities in the ontology. A SPARQL query example is depicted next:

```
SELECT ?x
WHERE
{
  ?x rdf:type :Employee
}
```

In this example, the *WHERE* clause contains a triple in which the first element is a variable (*?x*), the second element is the property *rdf:type*, and the third element is the class *Employee*. When the query is executed, variable *?x* is bound to those entities of the ontology related to *Employee* through the property *rdf:type*.

6.3   A Location and Context Based Access Control System

To illustrate our system, a scenario about large sports events will be used throughout this section [8]. This scenario includes multiple resources and actions, mobile users and resources, multiple ownership of resources, and collaborating organizations.

6.3.1   Scenario (Access Control Policies)

In large sports events, there are several organizations collaborating with each other. In this scenario, there are three organizations: *City Organizers*, *Police*, and *Medical Personnel*. A picture that depicts this scenario is shown in Fig. 6.3. The *City Organizers* are responsible for managing operations inside the venues where sports events are held. Operations include directing spectators to their seats, repairing equipment and providing services to athletes. The *Police* is responsible for safety and order, for checking bags and people at venues entry points and for monitoring sports events. The *Medical Personnel* is responsible for managing health care services and emergencies.

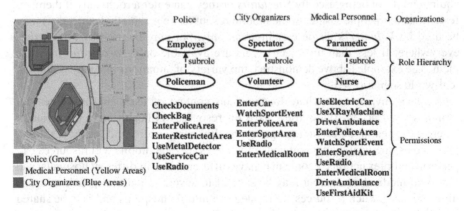

Fig. 6.3 Map of the areas owned by three organizations is shown on the *left-hand side*. Role hierarchies and permissions for each organization are depicted on the *right-hand side*. Roles are represented by *oval nodes* and role hierarchies by *dashed arrows*

Each organization has ownership over a set of resources. Resources may be either mobile or fixed in their positions. Mobile resources include cars and ambulances. Resources with fixed positions include areas inside the venues where sports events are held and equipment inside those areas. Each resource is associated with actions that can be performed on that resource. Following the RBAC terminology, we denote *<action, resource>* pairs as *permissions*. performed on the resources and that can be restricted to specific categories of people. For instance, *<Enter, PoliceArea>*, *<Drive, Ambulance>*, *<Use, FirstAidKit>* are permissions.

Each organization defines its own role hierarchy. Figure 6.3 depicts simple role hierarchies using ovals containing role names and dashed arrows. Examples of roles include *Spectator* and *Volunteer* belonging to *City Organizers*, *Employee* and *Policeman* belonging to *Police*, and *Paramedic* and *Nurse* belonging to *Medical Personnel*. Each role has permissions that allow the members of that role to perform their duties. Figure 6.3 depicts some permission names under the role hierarchies of each organization (e.g., *UseElectricCar*, *UseXRayMachine* under the *Medical Personnel* role hierarchy).

The assignment of users to roles is dynamic and depends on several conditions. For instance, people can be assigned the role *Volunteer* if they are inside the venues where sports events are held, but only during the time of the events and only if they display an appropriate certificate. The role of *Volunteer* can be further specialized in certain areas, where *Volunteers* must be also 21 years or older. As people move around and their location changes, their context changes as a result and therefore their roles also change. Roles are associated with different permissions inside different locations. For instance, *Volunteers* are allowed to enter special areas inside a *Stadium* and watch sports events for free, whereas they cannot watch sports events for free inside another *Stadium*.

Usage of resources depends on several conditions and therefore permissions on those resources may not always be available. For instance, *Volunteers* can use certain equipment if not being used by *Spectators* or they can enter a room only if there are fewer than ten people inside. Some mobile resources (e.g., small electric cars) may be used by *Volunteers* inside certain areas only, but may be used by *Paramedics* everywhere. If less than five electric cars are available only *Paramedics* may use them. The examples above demonstrate the variety of situations that may occur in a real-world scenario.

Another environment where location and context of users and resources determines what access rights users have on resources is health care, in particular emergency health care. In emergency health care, users and resources (e.g., paramedics, medical equipment and ambulances) can be mobile and resources can perform complex actions. People may have different roles in different hospitals, and access to medical equipment may be subject to several conditions, such as usage time and costs. Such resources are located in multiple hospitals and may be shared in collaborative scenarios that involve several hospitals.

6.3.2   A Context-Aware RBAC Model (Access Control Model)

Our scenario illustrates different kinds of access control policies where the assignment of users and permissions to roles depends on location and context of users and resources. We define *context* as the set of aspects of the environment, of the users or of resources relevant to access control. In location-based applications the context of users and resources may be variable. To be able to enforce these access control policies, an RBAC system (as represented in Fig. 6.1) must be able to represent and enforce constraints that can be grouped under the following two categories:

User-role assignment constraints These constraints are strictly connected to roles and specify how roles may be assigned to users depending on the users' context. They may include conditions on user location, environment conditions at that location, interaction with other users, and other conditions related to the user. To assign a user to the correct role, the enforcement mechanisms must be able to match the current context of the user against the conditions specified for each role.

Permission-role assignment constraints These constraints assign permissions to roles depending on the context of the resources on which those permissions are defined. For instance, the permission *DriveAmbulance* may be assigned to a *Paramedic* only if the ambulance is available or if it is at a specific location. More specifically, the access control model must be able to represent a wide range of resource-related conditions under which each permission may be assigned to roles. To assign the correct permissions to a role, the enforcement mechanisms must be able to match the current context of the resources against these conditions.

To address this problem, we extend RBAC with the notions of context and context constraint. A *context* C is a set of couples: $C = \{(a, v)\}$, where a is an attribute name associated with an RBAC entity and v is a value of that attribute. A *context constraint CC* is a set of couples: $CC = \{(a, a_c)\}$, where a is an attribute name and a_c is an *attributeconstraint* [8]. *Attribute constraints* define ranges of values and are defined as follows:

```
attributeconstraint ::= (attributeconstraint)
                       | RELATIONALOPERATOR constant
                       | NEGATION (attributeconstraint)
                       | attributeconstraint BINARYBOOLEANOPERATOR
                         attributeconstraint
```

where a constant can be of different types (e.g., string, number, boolean, area) and therefore the relational operator (e.g., $=$, $\leq$) is polymorphic in that it is able to compare different types (for example, $\leq$, when used for areas will be equivalent to geographic containment between two areas). Examples of attribute constraints include: $(\geq 10 \wedge \leq 18)$ and $(\leq Stadium)$. We say that a specific value v of an attribute *a satisfies* an attribute constraint a_c if v is inside the range of values defined by the attribute constraint and the attribute name in a_c is the same as the attribute name of v. Context C satisfies a context constraint CC if two conditions hold:

1. Every attribute name that appears in CC appears also in C. In other words, the context has all the attributes of the context constraint.
2. Every attribute value associated with an attribute name in C satisfies an attribute constraint in CC.

In our access control model, contexts and context constraints are associated with RBAC entities as follows:

Role context constraints These are pairs $<R,CC>$, where R is a role and CC is a context constraint. The context constraint CC specifies the conditions under which users may be assigned to role R. For instance, $<Volunteer,(certificate,=validated),(location,=Stadium),(time,=7pm)>$ is a role context constraint, representing the policy that "anybody whose certificate is *validated* and whose location is *Stadium* at *7pm* is a *Volunteer*." Role context constraints are defined by the security administrators and may be modified only by them.

User contexts These are pairs $<U,C>$, where U is a user, and C is a context. For instance, $<Alice,(certificate,=validated),(location,=pool)>$ is Alice's context. User contexts may vary as users move around. In particular, context attribute values may change (e.g., change of location) and attributes may be added to or removed from contexts. The user U is assigned to those roles whose context constraint is satisfied by the user's context.

Resource contexts These are pairs $<R,C>$ where R is a resource, and C is a context. For instance, $<ServiceCar,(location,=Parking),(NoPeople,=2)>$, is a resource context representing the current location and number of people inside a *ServiceCar*. the resource moves around.

Permission context constraints These are tuples $<P,V,CC,b>$, where P is a permission, V is a user or a role, CC is a context constraint, and b is a boolean constant. For instance, $<UseServiceCar,Volunteer(Alice),(NoPeople,\geq 1), false>$ is a permission context constraint. If the context of the resource associated with the permission P satisfies the context constraint CC, then P is enabled for V if b is true or is disabled for V if b is false.

To represent and enforce contexts and constraints and RBAC policies, we adopted the OWL language described in Sect. 6.2.2 [7, 8]. The choice of OWL is mainly motivated by its high expressive power. OWL contains several constructs (e.g., transitive, functional and symmetric properties, class composition) that can be used to represent complex relationships and complex role membership conditions in the constraints. In particular, the ability of OWL reasoners to *realize* OWL individuals depending on their property values makes OWL an ideal language to represent role context constraints and to enforce user-role assignment relations. In addition, OWL reasoners use only two reasoning features: realization and classification. Our framework uses these two features as a standard enforcement mechanism across organizations during collaboration.

6.3.3   OWL Representation and Enforcement (Enforcement Mechanisms)

In our framework, OWL is used to: (1) define *context ontologies* that represent users, resources, roles, and their *contexts*; (2) define *security ontologies* that represent RBAC policies. OWL reasoning is used for two purposes: (1) to assist security administrators during RBAC modeling of access control policies; (2) to enforce those access control policies.

6.3.3.1   Access Control Policies Modeling

OWL reasoning is used to derive an RBAC model given the entities of the domain and access control policies about those entities. To do this, security administrators are assisted by context and security ontologies, as depicted in Fig. 6.4 [6]. This figure shows a context ontology on the left-hand side. The context ontology contains OWL classes that describe categories of people (e.g., *Employee, Manager*) and resources (e.g., *Stadium*) and relationships between these categories. Figure 6.4 shows the OWL classes of a security ontology on the right-hand side. The security ontology contains classes that represent permissions (e.g., *EnterStadium*) and actions (e.g., *Enter*). These OWL classes are defined by security administrators who connect them with the classes in the context ontology. Figure 6.4 also shows an *RBAC ontology*.

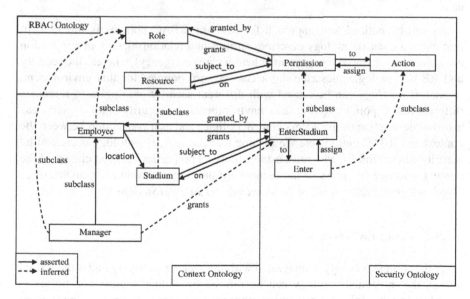

Fig. 6.4 Integration of context and security ontologies with reasoning. The properties that are represented by *solid arrows* are added by the security administrators, while those that are *dashed* are added by the reasoner

The RBAC ontology contains OWL classes that represent the RBAC entities and object properties connecting them. The names of these object properties are fixed; their definitions serve as a pattern used by the reasoner to classify classes of the context ontology as roles, permissions, resources, and actions. The definitions of these object properties are as follows [6]:

grants ⊂ *Role* × *Permission* The property *grants* connects a role with a permission. The inverse property of *grants* is called *granted_by*. If two OWL classes are connected by a *grants* and a *granted_by* property, then the first one is classified as a role by the reasoner and the second one is classified as a permission. For example, class *Employee* is classified by the reasoner as a role and class *EnterStadium* as a permission. The reasoner extends this type of classification to those OWL classes that are not directly connected to classes in the security ontology. For instance, class *Manager* is classified under Role since it is a subclass of *Employee*.

subject_to ⊂ *Resource* × *Permission* The property *subject_to* connects a resource with a permission. The inverse property of *subject_to* is called *on*. If two OWL classes are connected by these two properties, then the first one is classified by the reasoner as a resource and the second one as a permission. For example, *Stadium* is classified by the reasoner as a resource.

to ⊂ *Permission* × *Action* The property *to* connects a permission with an action. The inverse property of *to* is called *assign*. If two OWL classes are connected by these two properties, then the first one is classified by the reasoner as a permission and the second one as an action. For example, *Enter* is classified by the reasoner as an action.

As can be noticed, our approach for deriving RBAC policies is modular. In particular, a context ontology describes entities and relationships of an application environment (e.g., large sports event or health care emergency), whereas the security and RBAC ontologies describe the access control policies for that environment. A context ontology can be reused with different security ontologies to represent different RBAC policies for the same environment. The security ontology serves as a sort of description of the access control policies and as a connection between the context and RBAC ontologies. We assume that the security ontology is created by security administrators, who must establish object properties between classes of the context ontology and those of the security ontology. permission assignments (e.g., *Employee grants EnterStadium, Stadium subject_to EnterStadium*).

6.3.3.2   Policy Enforcement

After the context ontology is integrated with the security ontology under the RBAC model, the integrated ontology can be refined by adding context constraints to its roles. In what follows, we list our choices for modeling contexts and context constraints via OWL properties and values. To illustrate our modeling choices, we use examples from our scenario and use the Turtle notation to present OWL definitions.[1] In this notation, OWL definitions are represented by sets of triples. Each triple represents two entities of the ontology that are connected by a property. When several triples share the first element, then the first element appears only in the first triple. Ontology entities are preceded by prefixes, indicating their namespaces (e.g., owl, rdf, rdfs, or rbac). In particular, the *rbac* prefix is used for the OWL classes and properties of the RBAC ontology described in the previous paragraph.

Role representation. Roles are represented as OWL classes. To represent role context constraints, every OWL class representing a role is associated with properties whose values are restricted. These restrictions are expressed using the OWL constructs *owl:hasValue* and *owl:someValuesFrom*. For instance, to represent the role context constraint: $<Volunteer, (location, = Stadium), (certificate, = valid)>$, the class *Volunteer* is defined in OWL as follows:

```
:Volunteer rdf:type owl:Class ;
  owl:equivalentClass [ rdf:type owl:Restriction ;
                        owl:onProperty :location ;
                        owl:someValuesFrom :Stadium
                      ] ,
                      [ rdf:type owl:Restriction ;
                        owl:onProperty :certificate ;
                        owl:hasValue "valid"
                      ] ;
  rdfs:subClassOf rbac:Role.
```

Class *Volunteer* is associated with two restrictions on the properties *location* and *certificate*. The values of these properties have been restricted *Stadium* and *valid*, respectively. Class *Volunteer* is a subclass of class *Role*.

Session representation and user-role assignment. RBAC sessions are represented as OWL individuals. In particular, one OWL individual is created to represent each user session. That OWL individual is augmented with the properties and values of the user's context. As users move around and their contexts change, the property values of the corresponding OWL individuals are updated. The OWL reasoner realizes those OWL individuals into role classes depending on their property values. For instance, if *Alice* is located inside a *Stadium* and if her certificate has been validated, a session representing *Alice* is defined in OWL as follows:

```
:sessionAlice rdf:type owl:Thing ;
    :certificate "valid" ;
    :name "Alice" ;
    :location :Stadium .
```

This example shows the definition of the OWL individual *sessionAlice* to represent the RBAC session of *Alice*. This OWL individual has two datatype properties, *certificate* and *name*, and an object property, *location*. When it is created,

---

[1] http://www.w3.org/TeamSubmission/turtle/

sessionAlice is assigned to the most general OWL class (*owl:Thing*). The user-role assignment is enforced by the OWL reasoner, which realizes the session under the class *Volunteer*.

Resource representation. Resources are represented as OWL classes and individuals. A resource OWL class represents a set of similar resources. To represent the context of resources, the OWL individuals that represent them are associated with property values that are updated continuously. For instance, class *ServiceCar* and an individual of *ServiceCar*, *ServiceCar1*, are defined in OWL as follows:

```
:ServiceCar rdf:type owl:Class ;
            rdfs:subClassOf rbac:Resource ;
            rbac:subject_to UseServiceCar.

:ServiceCar1 rdf:type :ServiceCar ,
               owl :Thing ;
             :NoPeople "2";
             :availability "3pm" ;
             :location :ParkingLot ;
             rbac:subject_to :UseServiceCar1 .

rbac:subject_to rdf:type owl:ObjectProperty ;
            rdfs:range rbac:Permission ;
            rdfs:domain rbac:Resource .
```

Class *ServiceCar* is a subclass of class *Resource*. The property *subject_to* connects class *ServiceCar* with class *UseServiceCar*, which represents a permission. The definition of property *subject_to* is shown at the bottom of the example. *ServiceCar1* is defined as an OWL individual of type *ServiceCar*. The datatype properties *NoPeople, availability* and the object property *location* and their values represent the context of *ServiceCar1*. *UseServiceCar1* is an OWL individual that represents a permission (see below).

Permission representation. Permissions are represented as OWL classes and individuals. Each class represents a set of similar permissions. In the ontology, each permission individual is connected to the resource individual on which the permission is defined. For instance, to represent the class of permissions *UseServiceCar* and the permission individual *UseServiceCar1* that belongs to that class, the following OWL definition is used:

```
:UseServiceCar rdf:type owl:Class ;
            rdfs:subClassOf rbac:Permission .

:UseServiceCar1 rdf:type :UseServiceCar ,
                    owl :Thing ;
                rbac:on :ServiceCar1 .
```

Class *UseServiceCar* is a subclass of class *Permission*. *UseServiceCar1* is an OWL individual belonging to *UseServiceCar*. Property *rbac:on* is used to connect the permission individual *UseServiceCar1* with the resource individual *ServiceCar1* on which that permission is defined.

Permission-role assignment representation and enforcement. The permission-role assignment is represented and enforced at two different levels, which we call *class-level* and *individual-level*. The latter comprises the permission context constraints, which are used to enable or disable permissions depending on the current context of resources.

Class-level assignments represent general access control policies that are valid for all users of a role. For example, "*all Volunteers can use all service cars*" is a class-level assignment that is valid for all the individuals of the classes *Volunteer* and *UseServiceCar*. Class-level assignments are represented in OWL by constraining the range of the OWL object property *grants*, which connects role classes with permission classes. For example, the following restriction is used to specify that *Volunteers* can be granted the permissions represented by the class *UseServiceCar*:

```
:Volunteer rdf:type owl:Class ;
           rdfs:subClassOf rbac:Role ,
                    [ rdf:type owl:Restriction ;
                      owl:onProperty rbac:grants ;
                      owl:someValuesFrom :UseServiceCar
                    ] .

rbac:grants rdf:type owl:ObjectProperty ;
        rdfs:range rbac:Permission ;
        rdfs:domain rbac:Role ;
        owl:inverseOf rbac:granted_by .
```

The OWL construct *owl:someValuesFrom* is used to restrict the range of the object property *grants* to the class *UseServiceCar* if the domain of the property is the class *Volunteer*. This example also shows the definition of the property *grants* at the bottom. The enforcement of class-level assignments is performed in two steps:

1. User sessions are realized under a role class by the reasoner (e.g., *Alice is a Volunteer*).
2. Permission classes that are connected to the role classes via the object property *grants* are retrieved (e.g., *Volunteer grants UseServiceCar*). In practice, this means that the user session is granted all the permissions represented by the class *UseServiceCar*. Consequently, as a *Volunteer Alice* can use all service cars.

Individual-level assignments represent access control policies between single users or groups of users, and single permissions or groups of permissions. For instance, "*Volunteer Alice can use ServiceCar1 only if there is at least another person in the car*" is an access control policy about a single user, *Alice*, and a single resource, *ServiceCar1*. This policy can be modeled with the following permission context constraint: $<UseServiceCar1, Volunteer(Alice), (NoPeople, \geq 1), true>$.

Individual-level assignments are enforced using SPARQL queries [6]. In particular, SPARQL queries are used to verify conditions about resource individuals that OWL reasoners cannot detect. As mentioned in Sect. 6.2.2, these conditions may involve the join of property values of OWL individuals (e.g., co-location of two resources). We use the *ASK* construct of SPARQL for this purpose: it returns *true* if the context of a resource satisfies the conditions of the permission context constraint of the permissions defined on that resource (e.g., if the context of *ServiceCar1* satisfies the context constraint of *UseServiceCar1*). It returns *false* otherwise. Next, the logical AND between the result of the *ASK* query and the boolean value inside the corresponding permission context constraint is computed. If this final result is false, the permission is disabled, otherwise it is enabled. For instance, to check that the conditions exist for *Volunteer Alice* to use *ServiceCar1* the following SPARQL query is used:

```
ASK WHERE {
    ?x rdf:type :Volunteer
    ?x :name "Alice"
    :ServiceCar1 :NoPeople ?y
    FILTER (?y >= 1)
}
```

This query checks both the existence of a *Volunteer*, whose name is *Alice*, and whether or not the number of people inside *ServiceCar1* is greater than one. The FILTER construct restricts the focus of the query to those variables that satisfy the condition inside the parentheses. If the result of the query is false, the permission *UseServiceCar1* is removed from the list of permissions of the session representing *Alice*.

Relation between class-level and individual-level assignments. In our implementation, the set of permissions that can be disabled or enabled by individual-level assignments is contained in the set of permissions granted by class-level assignments. For instance, since *Alice* is a *Volunteer* she is granted all the permissions of the *UseServiceCar* class. Therefore, individual-level assignments for *Alice* are limited to the permissions belonging to the *UseServiceCar* class. This choice restricts our implementation within the scope of RBAC. Another possible choice is to use individual-level assignments to enable permissions that do not specifically belong to the roles of the session. In our view, this choice must be handled with care because, if abused, roles would become meaningless and the management of the access control policies would become more complex.

Representation of separation of duty constraints. Separation of duty (SoD) constraints specify that one or more roles cannot be assigned to the same user or activated for the same user at the same time. SoD constraints are modeled by the OWL construct *owl:disjointWith* that states that the sets of OWL individuals of two OWL classes are disjoint. If the constraint is violated, the reasoner raises an exception.

6.4   Interoperation Between RBAC Systems

In this section, we show how our OWL implementation supports interoperation between RBAC systems of different collaborating organizations. We use an example from the large sports events scenario. The *City Organizers* and the *Police*

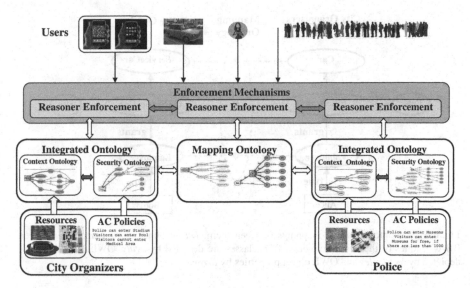

Fig. 6.5 A graphic representation of our interoperation model that depicts ontologies of different organizations and enforcement using reasoning

collaborate by sharing users and resources. For instance, a *Volunteer* belonging to *City Organizers* may enter areas under the jurisdiction of *Police*. Inside these areas, a *Volunteer* is assigned permissions belonging to *Police*, such as *UseRadio* and *EnterRestrictedArea*. To enable this type of collaboration among the two organizations, the RBAC system of the *City Organizers* should enforce the user-role assignment relation (e.g., assign the role *Volunteer* to *Alice*), whereas the RBAC system of the *Police* should enforce the permission-role assignment relation (e.g., allow Alice, who is a Volunteer, to use the radio).

Figure 6.5 shows our approach to enable interoperation between RBAC systems using OWL and reasoning. It depicts two organizations, which have ownership over sets of resources and related RBAC policies represented by OWL ontologies. *Mapping ontologies* are used to represent inter-organizational RBAC policies, which deal with extending roles, permissions and access to resources across organizations. These mapping ontologies contain OWL expressions that use classes of the ontologies that represent RBAC policies of the organizations. In particular, OWL expressions may represent equivalence or subsumption relationships between classes or combine several OWL classes into one OWL class. Thus, the mapping ontologies serve as a "bridge" between the RBAC policies of the organizations, by introducing new relationships and classes between the ontologies. Different mapping ontologies can be created for different collaborative situations. For instance, a mapping ontology for an emergency situation may be created in advance but only used when the emergency occurs.

Figure 6.5 also shows an abstraction of the enforcement mechanisms where several users request access to resources. OWL reasoning is used as a standard

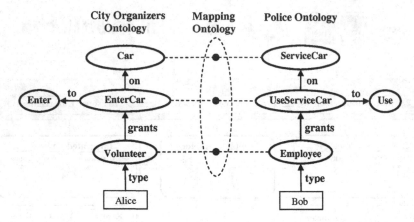

Fig. 6.6 Two ontologies and mappings between them. *Dashed lines* show mappings between RBAC entities of the two ontologies. OWL classes are depicted by *ovals*, OWL individuals are depicted by *rectangles* and OWL object properties by *arrows*

enforcement mechanism. In fact, the same OWL reasoner can be used on each ontology separately or on the mapping ontologies. In the former case, the reasoner enforces RBAC policies of each organization separately, while in the latter case it enforces the RBAC policies of all the organizations together with the inter-organizational RBAC policies.

OWL supports different kinds of mappings between OWL classes belonging to different ontologies, as shown in Fig. 6.6. This figure depicts two portions of the ontologies from two organizations and a schematic mapping ontology. The dashed lines depict mappings and OWL expressions between OWL classes of the two ontologies, which can be of different types:

Role mappings. Roles from different organizations may be mapped to one another under different relationships. In Fig. 6.6, these mappings are shown by the dashed lines between *Volunteer* and *Employee*. These relationships are expressed by OWL constructs, such as *owl:equivalentClass* and *rdfs:subClassOf*. The former states that two roles are equivalent and therefore the users of one role can be assigned to the other role. The latter states that one role is a subclass of another role, therefore extending the roles of users to the superclass. Other types of mappings can be defined using OWL class expressions in the mapping ontology. For instance, an OWL class can be created in the mapping ontology as the union of the OWL classes *Employee* and *Spectator*. Then, by stating that a *Volunteer* of the *City Organizers* is equivalent to this newly created class, users of role *Volunteer* can be assigned to both *Employee* and *Spectator*.

Permission mappings. Permissions from different organizations may be mapped similarly to roles. In this case, the permission-role assignment is extended from one permission to the corresponding mapped permission. For instance, permission *UseServiceCar* may be mapped through an equivalence OWL relationship to permission *EnterCar*. Consequently, users of role *Volunteer* are granted permission *UseServiceCar* via permission *EnterCar*, while users of role *Employee* are granted permission *EnterCar* via permission *UseServiceCar*. Another way to extend permissions to roles across organizations is to explicitly state in the mapping ontology that roles of one organization grant the permissions of the other organization (e.g., *Volunteer grants UseServiceCar*).

Resource mappings. Resources may be mapped to one another by using equivalence or subclass relationships. In this case, the type of access that exists on a resource of one organization will be extended to resources of the other organization. For instance, in Fig. 6.6 an equivalence relationship established between *ServiceCar* and *Car* extends the action *Use* to the resource *Car*. This type of extension may be desirable in situations in which specific actions of one organization may be executed on resources of other organizations.

To enforce the individual-level assignments in a collaborative scenario, SPARQL queries may be extended to users and roles of other organizations. Their execution is performed after the execution of the reasoner, as will be mentioned in the next section.

6.5 Implementation

We implemented our framework using the architecture shown in Fig. 6.7. The ontologies were written using the Protegé editor.[2] The OWL-API[3] and the Pellet reasoner[4] were used for loading/updating the ontologies and for reasoning. The steps that are followed for retrieving the set of allowed permissions of a user are listed next:

Step 1 (Access request). When a user requests access for the first time, the *New User Handling Module* is activated and a new OWL individual is created in the ontology to represent the user session. The attributes and values of the user context are attached to this OWL individual as OWL object or datatype properties. These attributes and values may either be sent by the user or be acquired periodically by the system. In our implementation, we use the first option.

Step 2 (Context update). The property values of OWL individuals representing resources, other users, or general attributes that are present in the ontology are updated by the *Attribute Handling Module*. The *Attribute Handling Module* is also responsible for the deletion of sessions.

---

[2]http://protege.stanford.edu/
[3]http://owlapi.sourceforge.net/
[4]http://clarkparsia.com/pellet/

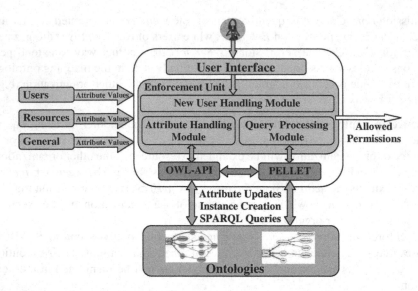

Fig. 6.7 Architecture of the access control system. The ontologies are updated with data sensed from the environment. The Pellet reasoner derives the permissions of the users, depending on the current state of the resources and on the property values of the users

Step 3 (Class-level assignment enforcement). The *Query Processing* module is used to obtain the roles to which the new user session belongs. During this step, the Pellet reasoner realizes the newly created OWL individual under one or more role classes using its property values. Permission classes associated with the user role classes are retrieved by following the restrictions on the object property *grants*. The permission individuals belonging to these permission classes are also retrieved. This set of individuals represents all the permissions granted to the user by the class-level assignments.

Step 4 (Individual-level assignment enforcement). The SPARQL queries associated with the permissions retrieved in the previous step are run. Depending on the result of the queries, some of those permissions may be either enabled or disabled. The scenario described in Sect. 6.3 was simulated in OWL using a variable number of RBAC roles, 33 RBAC resources, and one *Enter* or *Use* permission for each resource. Areas, users, roles, and permissions are displayed on a web page that uses the Google Maps API as depicted in Fig. 6.8. Each area is represented by a polygon. Users and mobile resources are represented by avatars. The location of users and resources is sent to a server together with other attribute values. These attribute values can either be manually selected or simulated automatically. The server runs the reasoner and sends the allowed permissions back to the web page, which displays them and changes the color of the unaccessible areas to red. The movement is simulated on the web page by dragging and dropping the avatars. As avatars move around, the set of allowed permissions changes.

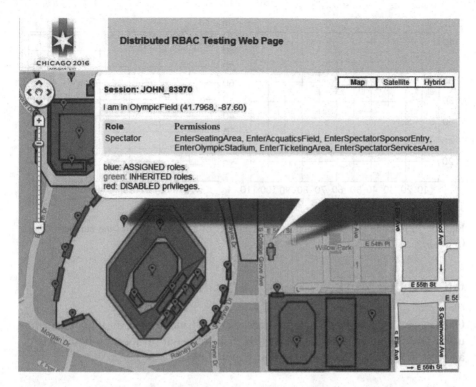

Fig. 6.8 Visual interface showing areas, users, permissions, and roles. Unaccessible areas are shown in *red*

The performance of the system on a Pentium Core 2 Duo machine with 2 GB of memory is shown in Fig. 6.9. This figure depicts graphs of the realization time as a function of the number of user sessions and as a function of the number of OWL classes representing RBAC roles. The x-axis on the left-hand side represents the number of user sessions, while the x-axis on the right-hand side represents the number of OWL classes representing roles. The time spent by the reasoner to realize a new user session is represented along the y-axis and is measured in seconds. When measuring realization time as a function of the number of user sessions, the number of role classes was kept constant at twenty classes. For example, the time spent by the reasoner to realize a new session is about seven seconds, when there are forty user sessions. When measuring realization time as a function of the number of role classes, there is only one OWL individual representing the user session in the ontology. For example, with 30 role classes and no other user sessions present, the reasoner takes around five seconds to realize a new user session. Performance starts degrading noticeably when 40–50 user sessions have been created in the ontology or if there are 50–60 roles present. Clearly more efficient reasoners are needed to handle large real world examples. In this regard, recent improvements to reasoners such as Pellet and FaCT++ are encouraging.

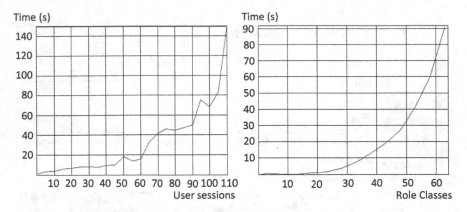

Fig. 6.9 Realization time as a function of the number of users (*left*) and of the number of roles (*right*)

6.6   Related Work

Our framework comprises both an access control model and its implementation using OWL. Our work is thus related both to research on location-based RBAC and research on implementation of RBAC with OWL.

6.6.1   *Location-Based RBAC*

Bertino et al. propose GEO-RBAC as an extension to RBAC [5]. Positions in GEO-RBAC can be real or logical: the former can be acquired through a location sensing technology, while the latter is a semantic notion that is defined at a higher level of abstraction (e.g., roads and addresses). A *location mapping function* computes logical positions from real positions. A *spatial role* in GEO-RBAC is defined by a name and a role extent, which is a set of logical positions. Users can request a role only if their position is logically contained within the role extent. GEO-RBAC focuses mainly on the association of roles with the spatial extents and on the activation and enabling of roles depending on a user's location. In our access control model, location is contained inside a wider range of attributes representing user contexts. In addition, we model the location of resources inside resource contexts, which determine if permissions are enabled or disabled, while GEO-RBAC deals exclusively with roles and user locations.

Aich et al. [1] and Ray and Toahchoodee [17] propose two similar approaches that associate spatio-temporal intervals with RBAC roles and resources. A spatio-temporal interval is composed of sets of points and of time intervals. The sets of points define logical locations. Roles can be assigned to users only if users are inside logical locations and only during the time intervals specified for those roles.

Permissions on a resource can be assigned to roles only if the resource is inside logical locations and only during the time intervals specified for the resource. Using our approach, context can be used to capture both space and time constraints.

6.6.2 OWL and RBAC

Several proposals have recently emerged that represent RBAC entities and assignment relations using OWL entities and relationships. Given a specific domain, the first step in building an RBAC system is the specification of the access control policies using the RBAC model: which resources must be protected, which permissions must be used, what roles are needed, and how are permissions assigned to roles and roles to users. After an RBAC model of the access control policies has been established, the next step is to map that model to an OWL ontology. There are several ways of mapping this model to OWL. In particular, the way RBAC concepts are represented affects the modeling choices of the relationships and of the constraints.

In ROWLBAC [13], two different ways are proposed for representing roles in OWL: roles as classes, where RBAC roles are represented as OWL classes, and roles as values, where RBAC roles are represented as OWL individuals. RBAC resources and actions are modeled as OWL classes. RBAC policies described by ontologies are enforced either by using OWL reasoning, if roles are represented as classes, or by using rules reasoning, if roles are represented as individuals. Our OWL implementation represents roles as OWL classes but uses a hybrid modeling choice for resources and permissions, in order to support fine-grained specifications of access control policies at class level and individual level. We also propose a simple way of combining the permissions that result from enforcement at both levels.

In Proteus, Toninelli et al. [21] define contexts as intermediaries between users and actions that the users can perform on resources. Contexts are defined as "any information that is useful for characterizing the state or the activity of an entity or the world in which this entity operates." Contexts, represented by OWL ontologies, include information about resources, users, and constraints on the actions that the users can perform. Our access control model stays within the bounds of RBAC, whereas Proteus does not use roles. Their work also uses the notion of context but in our work context is associated with RBAC entities and the whole model is subordinated to RBAC. In Proteus context is a first class concept, which is used to associate users to actions directly, whereas in our approach the representation of the access control policies is more modular and easier to manage because it relies on RBAC. This increased modularity also facilitates the representation of inter-organizational access control policies in collaborative scenarios.

Shafiq et al. propose a framework for integration of local RBAC policies into a global RBAC policy [20]. In their approach, permissions can be equivalent with permissions of other organizations. Their method of integration of RBAC policies establishes role mappings automatically, based on relationships between the corresponding permission sets. Our implementation using OWL allows us to represent more types of mappings. Furthermore, in our approach we also extend mappings to resources.

6.7   Challenges, Conclusions, and Future Directions

This chapter described an RBAC-based access control system that uses Semantic Web technologies to represent and enforce context-aware access control policies. We use Semantic Web languages and reasoners that are able to model a wide range of RBAC policies on top of a standard enforcement mechanism. This enforcement mechanism is also used to support interoperation between several access control systems during collaboration.

There remain several challenges to be addressed. One of the challenges is related to the realization time. One possible solution we plan to investigate is that of a modular ontology design. The main idea is to keep the size of the ontologies small enough to perform reasoning efficiently, while at the same time be able to represent complex interactions among the entities in the system. Another challenge is related to attribute values. In particular, we assume that the attribute values of users and resources are always accurate. This assumption may not be true in at least two cases: (1) location and data privacy of users must be protected, therefore an organization cannot acquire the exact location of a user; (2) location sensing technology provides an approximate location value. Each of these cases requires different solutions. In particular, to protect location and data privacy of users, a trusted third party may need to be introduced, which acts as an intermediary between users and organizations. To address location sensing technology limitations, approaches that incorporate uncertainty into the access control model may be useful [2].

From a modeling point of view, more complex types of collaboration between organizations can be investigated. Our access control model uses static mappings between classes in the ontologies. In complex collaborative scenarios, mappings may have a more dynamic nature; for instance, they may be established only after certain actions are performed by one of the organizations. They may also need to be constrained by complex conditions, which may or may not be expressible in OWL. We also assumed that there is complete trust between organizations. Therefore, the access control policies as represented by ontologies may be totally or partially revealed, so that the mappings can be established. In situations where the degree of trust does not allow for open mappings, privacy-preserving mapping techniques between access control ontologies may need to be investigated [9].

XACML is an XML-based framework designed for facilitating interoperation between access control systems [15]. XACML access requests consist of sets of attributes and values. XACML defines a *Policy Enforcement Point*, where access requests are received, and a *Policy Decision Point (PDP)*, where the security constraints are evaluated. In our access control system, the use of OWL reasoners is functionally similar to that of PDPs. We propose to investigate how our framework can be employed in an XACML framework. Recent work has addressed the use of the OWL language and reasoning in conjunction with XACML, but has dealt mainly with enforcing SoD constraints [12].

Acknowledgements This work was supported in part by NSF Awards IIS-0513553 and IIS-0812258. We would also like to thank Greg Jarzab for his contributions to the prototype implementation.

References

1. S. Aich, S. Mondal, S. Sural, and A. K. Majumdar. Role Based Access Control with Spatiotemporal Context for Mobile Applications. *Transactions on Computational Science IV: Special Issue on Security in Computing*, pages 177–199, 2009.
2. C. A. Ardagna, M. Cremonini, E. Damiani, S. D. C. di Vimercati, and P. Samarati. Supporting Location-based Conditions in Access Control Policies. In *Symposium on Information, Computer and Communications Security (ASIACCS)*, pages 212–222. ACM, 2006.
3. F. Baader, D. Calvanese, D. L. McGuinness, D. Nardi, and P. F. Patel-Schneider. *The Description Logic Handbook: Theory, Implementation, and Applications*. Cambridge University Press, Cambridge UK, 2003.
4. S. Bechhofer, F. van Harmelen, J. Hendler, I. Horrocks, D. L. McGuinness, P. F. Patel-Schneider, and L. A. Stein. OWL Web Ontology Language Reference. Technical report, World Wide Web Consortium, 2004. http://www.w3.org/TR/owl-ref/.
5. E. Bertino, B. Catania, M. L. Damiani, and P. Perlasca. GEO-RBAC: A Spatially Aware RBAC. In *Symposium on Access Control Models and Technologies (SACMAT)*, pages 29–37. ACM, 2005.
6. L. Cirio, I. F. Cruz, and R. Tamassia. A Role and Attribute Based Access Control System Using Semantic Web Technologies. In *International IFIP Workshop on Semantic Web and Web Semantics*, volume 4806 of *LNCS*, pages 1256–1266. Springer, 2007.
7. I. F. Cruz, R. Gjomemo, B. Lin, and M. . A Location Aware Role and Attribute Based Access Control System. In *ACM SIGSPATIAL International Conference on Advances in Geographic Information Systems (ACM GIS)*, pages 527–528. ACM, 2008.
8. I. F. Cruz, R. Gjomemo, B. Lin, and M. Orsini. A Constraint and Attribute Based Security Framework for Dynamic Role Assignment in Collaborative Environments. In *International Conference on Collaborative Computing: Networking, Applications and Worksharing (CollaborateCom)*, pages 1–18. Springer, 2008.
9. I. F. Cruz, R. Tamassia, and D. Yao. Privacy-Preserving Schema Matching Using Mutual Information. In *IFIP Conference on Data and Applications Security (DBSec)*, volume 4602 of *LNCS*, pages 93–94. Springer, 2007.
10. M. L. Damiani and E. Bertino. Access Control and Privacy in Location-Aware Services for Mobile Organizations. In *International Conference on Mobile Data Management (MDM)*, page 11. IEEE Computer Society, 2006.
11. D. Ferraiolo and R. Kuhn. Role-Based Access Control. In *NIST-NCSC National Computer Security Conference*, pages 554–563, 1992.
12. R. Ferrini and E. Bertino. Supporting RBAC with XACML+OWL. In *Symposium on Access Control Models and Technologies (SACMAT)*, pages 145–154. ACM, 2009.
13. T. W. Finin, A. Joshi, L. Kagal, J. Niu, R. S. Sandhu, W. H. Winsborough, and B. M. Thuraisingham. ROWLBAC: Representing Role Based Access Control in OWL. In *ACM Symposium on Access Control Models and Technologies (SACMAT)*, pages 73–82. ACM, 2008.
14. J. B. Joshi, R. Bhatti, E. Bertino, and A. Ghafoor. Access-Control Language for Multidomain Environments. *IEEE Internet Computing*, 8:40–50, 2004.

15. B. Parducci, H. Lockhart, R. Levinson, and J. B. Clark. OASIS eXtensible Access Control Markup Language (XACML) TC, 2005. http://www.oasis-open.org/committees/tc_home. php?wg_abbrev=xacml.
16. E. Prud'hommeaux and A. Seaborne. SPARQL Query Language for RDF. Technical report, World Wide Web Consortium, 2007. http://www.w3.org/TR/2007/WD-rdf-sparql-query-20070326/.
17. I. Ray and M. Toahchoodee. A Spatio-temporal Role-Based Access Control Model. In *Data and Applications Security XXI*, volume 4602 of *LNCS*, pages 211–226. Springer Berlin / Heidelberg, 2007.
18. R. S. Sandhu, E. J. Coyne, H. L. Feinstein, and C. E. Youman. Role-Based Access Control Models. *Computer*, 29(2):38–47, 1996.
19. R. S. Sandhu, D. F. Ferraiolo, and D. R. Kuhn. The NIST model for Role-based Access Control: Towards a Unified Standard. In *ACM Workshop on Role-Based Access Control*, pages 47–63, 2000.
20. B. Shafiq, J. B. D. Joshi, E. Bertino, and A. Ghafoor. Secure Interoperation in a Multidomain Environment Employing RBAC Policies. *IEEE Transactions on Knowledge and Data Engineering*, 17(11):1557–1577, 2005.
21. A. Toninelli, R. Montanari, L. Kagal, and O. Lassila. Proteus: A Semantic Context-Aware Adaptive Policy Model. In *International Workshop on Policies for Distributed Systems and Networks (POLICY)*, pages 129–140. IEEE Computer Society, 2007.
22. Vincent C. Hu and David F. Ferraiolo and D. Rick Kuhn. Assessment of Access Control Systems, 2006. http://csrc.nist.gov/publications/nistir/7316/NISTIR-7316.pdf.
23. C. Zhao, N. Heilili, S. Liu, and Z. Lin. Representation and Reasoning on RBAC: A Description Logic Approach. In *International Colloquium on Theoretical Aspects of Computing (ICTAC)*, volume 3722 of *LNCS*, pages 381–393. Springer, 2005.

Chapter 7
Topographic Mapping Data Semantics Through Data Conversion and Enhancement

Dalia Varanka, Jonathan Carter, E. Lynn Usery, and Thomas Shoberg

Abstract This paper presents research on the semantics of topographic data for triples and ontologies to blend the capabilities of the Semantic Web and *The National Map* of the U.S. Geological Survey. Automated conversion of relational topographic data of several geographic sample areas to the triple data model standard resulted in relatively poor semantic associations. Further research employed vocabularies of feature type and spatial relation terms. A user interface was designed to model the capture of non-standard terms relevant to public users and to map those terms to existing data models of *The National Map* through the use of ontology. Server access for the study area triple stores was made publicly available, illustrating how the development of linked data may transform institutional policies to open government data resources to the public. This paper presents these data conversion and research techniques that were tested as open linked data concepts leveraged through a user-centered interface and open USGS server access to the public.

7.1 Introduction

Since the 1990s institutional centralization of digital databases for domestic national topographic mapping in the United States has created greater rigidity and ambiguity of semantic meanings of landscape features. Analogue maps have a strong component of field verification, and when compiled and drafted from regional surveys, the feature types and names reflect localized interpretations more than when standardized into a central national database. As topographical data changed media from paper maps to digital databases, features became coded as segments of a data model where names and feature types were assigned a broadly unifying

D. Varanka (✉)
United States Geological Survey, Rolla, MO, USA
e-mail: dvaranka@usgs.gov

N. Ashish and A.P. Sheth (eds.), *Geospatial Semantics and the Semantic Web:*
Foundations, Algorithms, and Applications, Semantic Web and Beyond 12,
DOI 10.1007/978-1-4419-9446-2_7, © Springer Science+Business Media, LLC 2011

thematic domain, such as 'transportation,' and not as entities within the visual context of places on the map, adding to the loss of topographic semantic meaning. To build relevance and responsiveness to diverse public viewpoints, local and regional context can be restored to the twenty first-century version of national topographic mapping, *The National Map* of the USGS [1]. The aim of applying semantic technology to topographic data and mapping is to build this context by specifying meanings of and relations among geospatial features.

The concept of topography has variable semantic meaning. Historically, topography referred to the local scale of the environment, as Ptolemy meant it [2]. To others, topography may mean the domain of landforms or surface of the earth [3]. In more socially-oriented studies, topography is sometimes defined as the direct experience of the landscape as people move through the environment [4, 5]. Since most environmental experience is limited to walking or other slow forms of transportation, such experience would necessarily be local, but in recent times, time and space have become compressed and people have access to local experiences as technologically represented. Increasingly, local environmental experience does not include landforms, but rather the built environment. It is this socially-oriented topography that is used in this study.

The National Map includes the USGS geospatial and topographic mapping data and services with eight base data layers: transportation, structures, orthoimagery, hydrography, land cover, geographic names, boundaries, and elevation; and public domain access to these and other data through a Web portal, *The National Map* Viewer. *The National Map* is a collaborative effort built on partnerships and standards to improve and deliver topographic information for the nation at multiple scales and resolutions. The goal of *The National Map* is to become the nation's source for trusted, current, and integrated topographic information available online for a broad-range of uses. This goal and a policy of collaboration make *The National Map* compatible with the vision for the Semantic Web, a web of broadly linked data over the Internet [6].

The basic data units of the Semantic Web are triples of the Resource Description Framework (RDF), represented as two nodes connected by an edge. The data the triples are intended to convey require a vocabulary of terms whose meaning is represented by the resources of the data model; the subject (a node), a predicate (the edge), and the object (the second node). The predicates indicate relations between object/subject resources. Each resource carries with it a namespace, notation for vocabulary taking the form of a Universal Resource Identifier (URI), a string of characters used to identify the resource on the Internet [7]. Triples link along the nodes of their resources when their namespaces are identical. These linked data form the graph of the Semantic Web. The engineering of these triples involves RDF or Web Ontology Language (OWL) [8]. The standard for information queries applied to triples is the SPARQL Protocol and RDF Query Language (SPARQL) [9].

This basic technology was applied to data from *The National Map* for prototype conversions from relational tables to triples, to allow users to link with other graph data contributed over the Internet from across the world. This paper describes this work and is structured as follows. The next section provides background on landscape change and topographic information and the need for flexible data formatting, such as RDF triples. Section 7.3 details the explicit procedures for converting point and line data to RDF triples. Section 7.4 provides a description of topographic semantics of feature types and relations. Section 7.5 describes the need and availability for public access to the semantics of topographic data and the USGS solution of a publicly accessible server and database query endpoint. The final section draws conclusions from the work.

7.2 Language and Landscape Change

The currency of topographic data semantics for *The National Map* is expected to be found with contemporary data users by allowing them to contribute to the design of the vocabulary and data model from their own viewpoints and experiences. Concurrently, the USGS data triples provided for public use differ from similar programs of other nations or community-based linked-data projects because the data are intended to support the USGS science strategy [10]. In addition to its availability for individuals, organizations, and industry, the USGS provides scientifically-oriented geographic information for various applications needed by researchers, such as data analysis and modeling. The potential to integrate disparate data for complex systems analysis by leveraging the linked-data model of triples and graphs will potentially better serve national science missions by distributing the sources and costs of collecting that data. Linked data, however, requires rules and organization for effective use.

To engage the capabilities of other users over the Semantic Web, so that the data quality is readily describable and minimally redundant, *The National Map* must maintain a basic, consistent model to which users enhancements of the data would be added. The logical axioms governing the execution of user enhancements to and queries on these data are implemented through the design of ontologies. Ontology is often defined as the "explicit specification of a conceptualization" [11]. Topographic ontology modules for *The National Map* were designed and developed to reflect topographic science concepts related by classes and subclasses of features, particularly through the use of spatial relations as triple predicate resources and the use of logical inference along the graph data model. An upper-level, top-down topographic feature ontology was developed employing scientific knowledge based on instance-level, bottom-up topographic feature sets. A gazetteer of topographic features was integrated for the start of populating the ontology classes. Gazetteers commonly consist of a feature name, unique identifier, categorized feature type, and a reference to a point of geographic location (despite that many features are linear or areal; the point is sometimes used in GIS as the location of the feature name in mapping). The gazetteer complements the linkage of topographical ontology feature classes to *The National Map* data by specifying these identifying elements.

The features and events represented in the triples must combine to reflect various facets of spatially-explicit information sought by users and be available for information extraction using natural language-based terms based on SPARQL queries. A wide array of topographic feature terms including synonyms and related terms could best reach the regional and demographic character of a diverse nation. An approach to test this objective was developed as a prototype involving the automated learning of topographic landform terms that are not already stored in the database, and which do not require the user to select from a limited number of pre-defined feature types. Allowing access to a SPARQL endpoint, a URL on the web that implements the SPARQL protocol, was a new development in USGS data access policies that were stretched to engage public data use that modifies the database.

Three main topics discussed in the following sections are the conversion of standard data, the design of new triple terms and topographic science ontologies, and the involvement of users in the development of these semantic technologies through the use of an interface and a gazetteer. Some implications of such developments for open data management policies is also noted. The pilot research projects undertaken to address the question of national topographic data semantics are not yet integrated into the operational constraints of USGS production practices. The concepts described in this paper explore the forms *The National Map* could or may take as linked data on the Semantic Web.

7.3   Geospatial Point and Vector Data Conversion to RDF

In the trial conversion of *The National Map* data sets to RDF, sample sets of relational topographic data were converted using open source tools. The objectives for this conversion were that the data be easy to use, contains only correct relationships (no identical URIs referencing two distinct objects or predicates with the wrong subject), and not lose any of the information found in the databases.

Point data from a gazetteer and GIS vector data (hydrography, transportation, structures, and boundaries) were converted using sample data sets originally developed as ESRI Arc/GIS shapefiles. The six watersheds and three urban are as represented in the data samples are: Pomme de Terre, MO; Upper Suwannee, GA-FL; Lower Prairie Dog Town Fork of the Red, TX; Lower Beaver, UT; South Branch of the Potomac, WV; and the Piceance-Yellow, CO; and Atlanta, GA; St. Louis, MO; and New Haven, CT. The representation of geometric shape through location coordinates was considered to be a particular challenge. Data conversion of *The National Map* data in the raster data model (elevation, land cover, images), is not presently developed, but methods for extracting features from these data and building the semantics around the features are in progress.

Fig. 7.1 A section of the GNIS database being converted to RDF

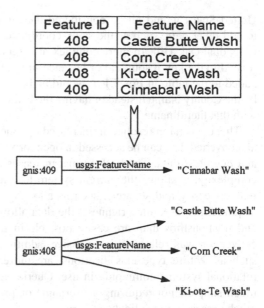

Feature ID	Feature Name
408	Castle Butte Wash
408	Corn Creek
408	Ki-ote-Te Wash
409	Cinnabar Wash

7.3.1 Converting Point Data

The Geographic Names Information System (GNIS) is a gazetteer of the U.S. Board on Geographic Names (USBGN), a partner of the USGS [12]. A custom program was written to convert subsets of GNIS tables, exported as a.csv (character separated value) into an RDF model. A GNIS namespace was assigned for the GNIS data and a USGS namespace was derived for this project data that are published by the USGS or described by its published standards. An example depicting a small section of the GNIS database before and after conversion is shown in Fig. 7.1.

In this example, the prefix "gnis:" represents http://usgs.gov/rdf/gnis/featureID/#. This means, for example, gnis:409 is the object identified by the URI http://usgs.gov/rdf/gnis/featureID/409. In this case, although it is a valid URL, the URI is strictly used as a globally unique identifier. The prefix "usgs:" represents http://usgs.gov/rdf/gnis/#; therefore, usgs:FeatureName is defined as the relation usgs.gov uses for GNIS when describing the feature name of a feature in RDF. When converting data from small, simple databases such as the example shown in the table in Fig. 7.1, a resource is formed by the key of the database row, in this case the Feature ID, and the other column headers are renamed using RDF naming conventions and given a URI to represent the relation. For each item in each row, a relation is formed between the key resource and the value of the column using the renamed column header as the relation name and making the value a resource or literal value as appropriate for the data in the column.

It usually is best to assign only items that have a universally constant meaning to resources and other values literals. For instance, if the Feature Name were a resource for only the All Names table in GNIS, all 220 instances of Oak Hill throughout the

United States would all be considered the same place. This makes it impossible to add other data, such as the state, coordinates, and feature type to identify each individual Oak Hill. In more complex data sets, not all of the information will be related to the key resource. For instance, if the County Name and County ID were added to the table in Fig. 7.1, it would be best to have the County ID be described by the County Name instead of having the Feature ID described as being in a county with that literal name.

This conversion resulted in increased openness compared to the relational data; all converted data can be accessed in open formats, such as text tables, RDF models, and plain-text data. The RDF format increases availability and decreases program complexity; it allows the current system to remain unchanged while adding new features easily and securely, such as allowing direct queries. The converted data acquired increased data richness; the data allow more complex data classification and relationships than are easily possible in traditional databases. Although the conversion was relatively inexpensive, addition to and enhancement of the data, such as feature type classifications, are more costly than the relatively simple relational systems commonly in use. Queries and accessing the database became much simpler, not requiring a username or password to access the server, or as much processing to validate a query.

Conversion is slower than table-based databases on typical server set-ups. It could take as long as 3 days to convert the entire GNIS database on a moderately powerful desktop from a group of pipe-delineated text files exported from a typical database into an RDF graph stored, for example, in a Jena Tuple Database triple store [13, 14]. The conversion time increases approximately as the square of the volume of data to be converted. Other datasets from *The National Map*, such as the National Hydrography Dataset (NHD), would take many days to months to convert to RDF, thus a faster conversion process must be realized. Possible solutions to make the conversion process more efficient include a parallel conversion algorithm for a Beowulf cluster. The approach is to parallelize the conversion and distribute the workload on a Beowulf cluster, a multi-core processor, or multiple machines on a network, to reduce the time necessary to convert these data.

7.3.2   Converting Vector Data

The primary challenge in converting geospatial vector data is the geometric representation of shape and location that is more complex than simply a point. The point data of a gazetteer is able to reuse well-known terms, such as the W3C's terms for latitude and longitude [15]. Geographical names for topographical features in *The National Map* are embedded in the vector data files.

These vector data files can be isomorphically represented by the Geography Markup Language (GML) [16]. To preserve the vector data in a queryable form in RDF, the entire feature member string (the GML representation of a single entity, for instance a line or area as well as its attribute data) is stored and the attribute data are extracted and converted to RDF. This approach allows the vector data to be queried normally while still allowing GML representations of the results set to be recombined into a valid GML document able to be processed and displayed using any GIS software. The only major trade-off using this set up is a minor increase in required storage space.

The semantic content of the converted data is identical to the original data. Semantic attributes in established GIS and geospatial data technologies are often manually entered. Even if extensive attributes are attached to the data files, these properties are not easily shared between data sets. These technical limitations have prevented the easy sharing of attributes between data. Considerations for expanding the range of semantic properties of data involve improving the usability of the data by others and the ability to interlink the data with the larger semantic community by following established conventions, such as the Linked Data [17] guidelines.

7.4   Topographic Data Semantics

Though the conversion of relational data to semantic triple data is not a new technical challenge, enriching the semantic meaning of the topographical data requires careful conceptualization. In the conversion of relational data, semantic meaning consists mainly of the meaning of the term itself if metadata are lacking or unreferenced, and the column heading that forms the predicate of the triple. Greater semantic meaning can be associated with triple data through the use of namespace definitions, ontology files, and data inference. Ontology in computer science ranges in the degree of the subject matter conceptualization and formalization in a system, but a basic practice of ontology design is a compiled vocabulary of terms with a description of their meaning [18]. In this study, standard vocabulary terms and definitions found in on-line sources provide a base which the public can query and enrich with new terms. New terms enrich the controlled vocabulary to make the data more responsive to users, if compiled in an organized way. Ontologies can moderate the structure of non-standard terms to be reliably incorporated into the database [19].

A project resembling Semantic Web ontology development was attempted by the USGS in the 1980s and resulted in a topographic feature type taxonomy with some spatial relations [20]. The data model was never executed as an application schema, but the taxonomy of this and the other USGS-related projects contributed toward the taxonomy of topographic feature types for this study. Most of the terms to be used as ontology classes and subclasses are based on standard topographic mapping data glossaries developed by USGS and its partners. These are the Digital Line Graph (DLG) and National Hydrography Dataset (NHD), the Spatial Data Transfer Standard (SDTS), and GNIS feature lists [21–24]. The DLG and NHD lists were based on the feature types that were compiled from years of repeated field validation over the United States landscape, and were interpreted by topographers as basic feature types that would be cognitively easy to recognize. Such feature types represent basic cognitive object categories and have a greater chance of having a widely-recognizable meaning [26, 26]. The SDTS feature list was developed with partners and has a wider scope of included features than the DLG. Features from the SDTS standard selected for this study include some coastal features, but excludes others that, as an international standard, are inappropriate to the United States interior, but used for cross-cultural ontology research [27]. The GNIS list originally was compiled from feature types taken from USGS topographical maps, but has added partner and volunteer contributions since 1987.

Though these standard terms have definitions, their invariant meanings may not be adequately captured and could impede interoperability, but most terms are basic and commonly used concepts within the shared sphere of their users, and offer an undetermined level of semantic clarity [28]. Also, these lists are not a comprehensive inventory of landscape features of the United States. Some feature classes were added to better complete the topographic ontology discussed in the next section and to insert concepts that are relevant to the feature type definitions and their associations to other classes. The use of designed feature codes, such as those used by the Federal Geographic Data Committee (FGDC), was avoided because they function as object classes with a certain degree of ambiguity to accommodate various meanings of data types for diverse users and applications.

Topographic/geospatial triples require spatial relations. Some solutions for spatial relation predicates and the related problem of spatial location are implemented within Internet-based projects [29–31]. A predominant source for spatial relation standards is the Open Geospatial Consortium (OGC) spatial relations (operators) standards, also accepted as International Organization for Standardization (ISO) 19125 – Simple Features Access [32]. Although the work of the USGS complies with OGC standards, a study of USGS standard feature glossary verbs and spatial prepositions seeks to identify basic terms indicating spatial descriptors, relations, and processes used for landscape modeling, such as 'used,' 'caused,' 'flows,' or 'removed' (Table 7.1) [33].

In addition to semantics used by the USGS, other standard vocabularies are employed, such as RDF, OWL, and the Simple Knowledge Organizing System (SKOS) [34].

7.4.1   The USGS Topographic Science Ontology Modules

Conceptual ontologies of topographic science were developed from USGS standards. The main ontology, called Topography, is a domain ontology of the subject matter of its name [35]. Topography consists of six modules consisting of topographic categorizations [36]. Relations among themes of the topography ontology reflect a general building or layering nature of topography (Fig. 7.2). Certain sub-themes help shape the characteristics of others. For example, terrain can be considered to generally direct the flow of surface water, and the characters of terrain and surface water have strong determinate effects on ecological regimes. Some

Table 7.1 Part of the analytical table for verb/spatial preposition analysis of GNIS features

Used		
Canal (manmade waterway)	usedBY	Watercraft drainage, irrigation, mining, or water power (ditch, lateal)
Channel (linear deep part of a body of water through which the main volume of water flows)	usedAS	A route for watercraft (passage, reach, strait, thoroughfare, throughfare)
School (building or group of buildings)	usedAS	As an institution for study, teaching, and learning (academy, college, high school, university)
Well (manmade shaft or hole in the Earth's surface)	usedTo	Obtain fluid or gaseous materials
Church (building)	usedFOR	Religious worship (chapel, mosque, synagogue, tabernacle, temple)
Military (place or facility)	usedFOR	Various aspects of or relating to military activity
Post office (an official facility of the U.S. Postal Service)	usedFOR	Processing and distributing mail and other postal material
Tower (a manmade structure, higher than its diameter)	usedFOR	Observation, storage, or electronic transmission
Airport (manmade facility)	usedFOR	Aircraft (airfield, airstrip, landing field, landing strip)

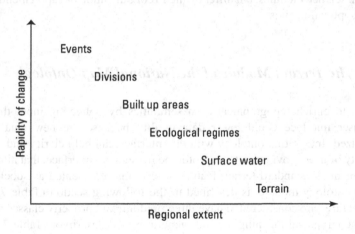

Fig. 7.2 Conceptual Model of a topographic ontology with six thematic modules

geographic characteristics were recognized to influence topography, though these are not themselves considered topographic; examples are "latitude" or "region." These were added as properties or as domains on the range of values. The thematic modules are considered to vary in the rapidity of their temporal change. For example, the temporal change of the six modules generally grows finer in the range between terrain and events. Those domains that change less rapidly tend to have greater regional extent.

Though the six classification headings may appear to be divided between natural (terrain, surface water, and ecological regime) and human-induced (built-up areas, divisions, and events) themes, none of these modules differentiate between 'natural' or 'artificial' features because of semantic complexity, sometimes because of their function. For example, if terrain or surface water features were modified artificially, a complex feature would result, such as 'mine' or 'flood zone' [37]. Each of the elements of complex features appear under a basic domain concept at the super-class heading with spatial and attribute relations to each other within their context.

Because topographic data primarily serve as a base for diverse manipulation and development by the public and scientists, no specific application is determined to drive the design of *The National Map* data. For this reason, the ontology modules or complex feature ontologies based on *The National Map* most closely approximate the design of reusable ontology design patterns [38]. These ontologies are semi-formal ontologies, described by Sheth and Ramakrishnan [39] as "...those that do not claim formal semantics and/or are populated with partial or incomplete knowledge." Topography exists as basic taxonomies in.owl files and further formalism is manually being designed based on linguistic and spatial semantic transformations. In addition to representing 'real world' topography, the subject domain modules are intended to be applied toward the development of task ontologies, such as topographic mapping [40]. Task ontologies will share common links with science modules, but differ by their reorganization or supplementation for specific application aims.

7.4.2   The Terrain Module of the National Map Ontology

Attempts to enrich topographical term semantics by collecting input through a public user interface could lower data quality because the new data are not standardized. Integrating ontology with the interface can help clarify and organize the quality of user-provided data. A prototype for such an interface, that allows user enhancements to standard terrain feature type terms represented as subclasses of a terrain ontology module, is described in the following section (Table 7.2). The terrain features are considered to have three predominant property classes referring to their geographical meaning: locator, generator, and descriptor (Table 7.3), with appropriate subclasses and definitions (not shown in the table). In addition to coordinate geometry drawn from *The National Map* data files, locator classes involve topological spatial relations. Generator properties refer to the prevailing physical environmental conditions and larger topographical context acting on the development of the landform, resulting in a descriptor. Shape, for example, is considered to be a descriptor.

Each of the pairs of properties is an inverse property and has a domain and range of subclasses. These properties and their associated axioms will help categorize non-standard feature type terms in the user interface.

Table 7.2 Standard terms for terrain features and feature type subclasses for ontology terrain module

Aeolian	Delta	Island	Plateau
Arch	Dish	Island cluster	Quicksand
Bar	Divide	Isthmus	Reef
Basin	Drainage basin	Karst	Ridge
Beach	Dunes	Lava	Ridge line
Bench	Fault	Mineral pile	Salt pan
Cape	Floodplain	Moraine	Shaft
Catchment	Fracture	Mount	Sink solution
Cave	Fumarole	Mountain range	chimneys
Chimney	Gap	Peak	Summit
Cirque	Glacial	Peneplain	Talus
Cliff	Ground surface	Peninsula	Terrace
Coast	Hill	Pinnacle	Valley
Crater	Incline	Plain	Volcano

Table 7.3 Ontology module properties as codes for triples, with no relation drawn between locators and descriptors

	Objects			
Subjects	F	L	G	D
FeatureType – F	1	2	3	4
Locator – L	5	1	5	n/a
Generator – G	5	2	1	6
Descriptor – D	7	n/a	3	1
Property codes				
owl:sameAs	1			
locatedAt	2	locationOf		5
generatedBy	3	generates		6
describedAs	4	describes		7

7.4.3 Compiling Gazetteer Feature Type Term Candidates

The standard format consisting of a feature name, coordinate position, category type, and a unique feature identifier can facilitate simple semantic queries such as 'where is' or 'what is.' In addition to not supporting complex queries, traditional gazetteers like the GNIS do not contain much local and vernacular topographical information. User-enrichment of the standard terms could possibly be used to augment the gazetteer and allow more complex queries through the use of the ontology.

To enrich the available feature terms within the gazetteer, a technique is needed to allow users to enter terms that will be organized and made available to other users. Public-input techniques characteristically freely provide place names and coordinate locations, but not the classification of these features into categories or classes [41].

An ontology benefits the input and the institutional database by providing structure that can draw on the capabilities of semantic technology and can help prevent some common errors of crowd-sourced information, thereby building trust in the data. The feature type categorization for publicly-provided information for *The National Map* is based on the standard vocabularies discussed above, with the new terms added in a systematic way to enhance the controlled standard. Queries for non-standard terms could be accessible through inference once multiple graphs are linked.

The advantage of using a gazetteer to interface with data users instead of users interfacing directly with the geospatial database itself is that the gazetteer allows the user to search with a place or feature name, such as "Grand Canyon," or feature types, such as "rivers." The gazetteer informs the user of the type of feature the name refers to, but since the place or feature is located as a point, the gazetteer does not map the feature as a geometric entity. For this reason, a gazetteer may be more complete or extensive than a GIS dataset in number of features, since the focus is only on the collection of names and their categorization and not the more expensive project of mapping. In GIS, the inclusion of names is linked to the collection of a feature in the dataset. Also, gazetteers are more flexible than feature data sets because the data files are smaller and more easily manipulated. Names linked with discrete feature objects in GIS data, however, offer the advantage of displaying additional attribution, such as the feature's length or extent. When a user queries a feature name or type though the gazetteer, the query can link through the gazetteer to the data through a feature identification number, or its coordinates. As alternative names or synonymous feature types are added to the gazetteer, the new terms enrich the access to the data without confusing the feature identification number or location. The feature terms used as classes in the ontology are also the labels of the triple instances, with the addition of spatial locations. These instances are derived from the GNIS.

7.4.4   Identifying New Feature Types

A prototype interface for querying a non-standard term in the Terrain module begins with a feature type term entered by the user, as shown in Fig. 7.3. The term is compared to the feature type list, assuming the feature type name is spelled correctly or could be compared to a typographical error checker. This prototype uses GNIS. If the user's feature type is unlisted with the gazetteer, the feature name is entered in an Internet dictionary and statements describing the feature term are collected and sorted. Ancillary information such as URL links is deleted. The remaining terms in the definitions are compared to the gazetteer feature list. Any that match appear on the interface to the user, along with an indication of the frequency the gazetteer term matches the same term in multiple definitions.

Preliminary results indicate that the most frequent gazetteer match to a user's term often is a synonym. Additional matches near the top of the list are related to

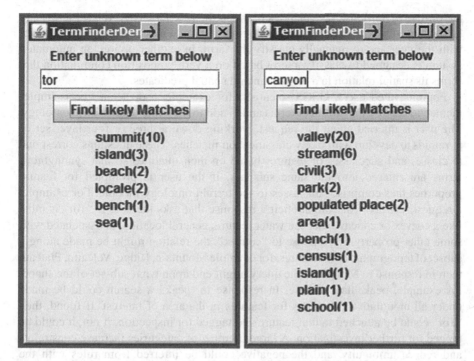

Fig. 7.3 User interface to search non-standard terms, and results for specific term searches

the users' query term; an implied preposition precedes the term [42]. In the example shown in Fig. 7.3, a 'tor' is a hill with rocks at the summit. Initially, related or synonymous terms appearing in the on-line definitions, such as 'hill' or 'rocks,' may be missed by the interface because they do not appear in the gazetteer. With time, the additional of new terms to the gazetteer will resolve that omission. 'Summit' is the most frequent association in this example; it has a descriptive role and a relation as part of a tor. The next most frequent term associated with tor in the example is island, which can be explained because a tor can be an island. The least well associated terms after that have remote relations to or roles for a tor and would be omitted. In the second example, using 'canyon,' the first term (valley) is a synonym for a canyon and the second term (stream) is the generating process of a canyon. In both cases, the first two terms fit into the topography ontology schema, but require sorting to meaningfully enhance the ontology.

7.4.5   Organizing New Feature Types into an Ontology

To populate the gazetteer feature list, the new feature term is classified in the database via the ontology based on the relation of the term to GNIS matches in the on-line definitions. The relations between the new terms and related terms can be developed by optional approaches. In one option, a motivated user can interact with the interface to manually classify the term. In another option, an automated system is designed to classify a term based on a verb/preposition combination that forms its spatial relation in a list of standard spatial predicates.

For motivated users, USGS scientists, for example, a decision tree is implemented to acquire the gazetteer terms and to add to the triple database via ontology. The user is queried about the request, working down a three or four layer set of pyramids to develop with likely classification matches. These questions correspond to classes and subclasses of features based on their qualities so that synonymous terms are entered into the same subclass, if the user is prompted for feature properties that comprise subclasses to the terrain ontology module. For example, a request such as "Tor" might elicit a response that asks the user if "Tor" is most like a convex or concave shape, a water feature, general location, or associated with some other property. In response to "convex", the relation might be made along a subset of topographic high features, for example Mountain, Ridge, Volcano, Plateau, then in response to Mountain, the query might end upon what sub-set of the shape, for example, peak, flat, or base. In response to 'peak,' a search could be made under all mountain top features for features in the area of interest. If found, then "Tor" could be attached to that feature and tagged for inspection, if not, it could be tagged for further investigation. A larger number of categories increase constraints and reduce ambiguity, and the negative could be inferred from rules with the term 'not.'

Automated approaches toward a user-enhanced vocabulary would be to compare feature classes to word/string patterns in readable sources. For example, spatial relation terms based on verbs and spatial prepositions were identified for triple predicates for each of the standard feature types listed by GNIS (Table 7.1). The subjects and objects of the GNIS glossary are classified in the topography ontology. If the predicate associated with subjects and objects in the GNIS glossary matched the predicate within the string between the subject and object in the on-line definition, the same relation would be assigned to the term associated with the predicate. An approach based on spatial verb/predicate associations shows promise, but needs more research.

Statistics on the number of collection occurrences that match existing classes indicate the reproducibility of a term, as well as indicate relatively unused terms. Saved feature type terms are possible gazetteer candidates when saved past a threshold number. Prioritizing the frequently used terms, together with the new terms, eventually result in a compiled set of terms in commonly current usage, particularly by the system users. The prototype indicates the potential to develop computational methods for developing vocabularies that are relevant and responsive to diverse data users.

7.5   Institutional and Public Data Interaction

The transition of data to digital form since the 1980s resulted in centralized government databases that are most often kept closed because of requirements for security that could be threatened by unauthorized access. Only limited internal access to national mapping data was allowed. Internet developments in social networking and linked data fundamentally question centralization and call for open data access. Internet culture likewise calls for a movement toward open government, but data stewards and administrators remain mindful of security threats to the tested and trustworthy quality of technologically sophisticated data representation.

As a response to greater expectations for public interaction, researchers in support of *The National Map* maintain a server separated from centralized internal networks and linked to a university network for research purposes. Data and project materials are freely shared for hybrid research and public use. Institutional support was built by informing administrators about open access activities in detail.

Several potential levels of interaction are possible via html links: downloading files or read-only access; server read/write access with credentials verification; or uploading new work to the government server. This work is isolated in a 'Jail,' or restricted space. A SPARQL endpoint, Open Virtuoso, was installed with a graphical interface for query validation. This software was selected because it easily accommodates different scales of data, maintains an open process, threads for parallel processing, and is well-known and familiar with open source software users.

The success of the expansion of the database, gazetteer, and ontology semantics, especially through the use of user enhancements, will depend on the volume and type of input and their management. Input entered by individual users, such as those that could be invited by publishing an article or solicited through discussion lists for teachers, naturalists, outdoor recreationalists, or geography colleagues, are a clean source of input with little or no potential for semantic error. Manual entry of individual terms could build a vocabulary specific to users of *The National Map*. For further research, operations that process passages of text could generate extensive lists with statistical trend information.

7.6   Conclusions

National topographic information continues to need to reflect regional and demographic diversity and greater integration and retrieval of geographic information. Topographic features, represented as free and trusted data representations, carry with them semantic schemas that make them readily useable. Converting national topographic mapping data to triple formats as linked data offers topographic data users the advantages of semantic technology and results in available data to populate the Semantic Web. Some semantic concepts relating to the data don't closely match the perspectives of an extensive public, particularly of non-expert users.

To address this weakness, topographic vocabularies are modified through the use of standard glossaries, ontology modules, and a gazetteer. Increased interactivity and community involvement are enabling the creation of controlled vocabularies with greater specificity and relevance to users. The terms are valid because they are part of the experience of the U.S. landscape, and carry linguistic and conceptual commonality as reflected through the frequency and consistency of their use.

References

1. U.S. Geological Survey (2010) The National Map. http://nationalmap.gov. Accessed July 14, 2010
2. Dilke, OAW (1987) The Culmination of Greek Cartography in Ptolemy. In: Harley JB, Woodward D (eds) The History of Cartography, Volume 1, Cartography in Prehistoric, Ancient, and Medieval Europe and the Mediterranean, University of Chicago Press, Chicago
3. Mark, DM, Smith B (2004) A science of topography: From qualitative ontology to digital representation. In: Bishop MP, Shroder JF Jr (eds) Geographic Information Science and Mountain Geomorphology, Springer, Berlin, Germany
4. Curry, MR (2006) Lugares publicos e practices privadas [Private practices and public places]. In: Sarmento J, de Azevedo AF, Pimento JR (eds) Ensauos de geografia cultural, Livraria Editora Figueirinhas, Lisboa
5. Leatherbarrow D (2004) Topographical stories, studies in landscape and architecture. University of Pennsylania Press, Philadelphia
6. W3C (2010) Semantic Web. http://www.w3.org/standards/semanticweb/. Accessed July 14, 2010
7. Mealling M, Denenberg R (eds) (2002) Report from the Joint W3C/IETF URI Planning Interest Group: Uniform Resource Identifiers (URIs), URLs, and Uniform Resource Names (URNs): Clarifications and Recommendations. Request for Comments 3305. The Internet Society, Reston, Virginia
8. W3C (2004) RDF Primer. http://www.w3.org/TR/2004/REC-rdf-primer-20040210. accessed July 14, 2010
9. W3C (2008) Semantic Web. http://www.w3.org/TR/rdf-sparql-query/. Accessed July 14, 2010
10. US Geological Survey (2007) Facing tomorrow's challenges—US Geological Survey science in the decade 2007–2017, US Geological Survey Circular 1309. US Geological Survey, Reston, Virginia
11. Gruber TR (1995) Toward principles for the design of ontologies used for knowledge sharing. International Journal of Human-Computer Studies 43 (5–6): 907–928
12. US Board on Geographic Names (2010a) Geographic Names Information System (GNIS). U.S. Geological Survey, http://geonames.usgs.gov/domestic/index.html. Accessed August 30, 2010
13. Jena (2010) TDB. http://openjena.org/wiki/TDB. Accessed August 30, 2010
14. JPPF (2010) http://www.jppf.org/. Accessed August 30, 2010
15. W3C (2003) Basic Geo (WGS84 lat/long) Vocabulary. http://www.w3.org/2003/01/geo/. Accessed July 22, 2010
16. Portele C (2007) OpenGIS® Geography Markup Language (GML) Encoding Standard. OGC 07–036, Version 3.2.1. Open Geospatial Consortium Inc, Wayland, Mass
17. Linked Data (2010) Linked Data – Connect Distributed Data across the Web. http://linkeddata.org/. Accessed July 22, 2010
18. Uschold M, Gruninger M (1996) Ontologies: Principles, methods and application. Knowledge Engineering Review 11: 93–155

19. Gruber T (2007) Ontology of Folksonomy: A Mash-up of Apples and Oranges. International Journal on Semantic Web & Information Systems 3 (2) http://tomgruber.org/writing/ontology-of-folksonomy.htm. Accessed on July 27, 2010
20. Guptil SC, Boyko KJ, Domarat, MA, Fegeas FG, Rossmeissl HJ, Usery EL (1990) An enhanced digital line graph design. US Geological Survey Circular 1048, Reston, Virginia
21. US Geological Survey (2001) Digital Line Graph Standards. http://rockyweb.cr.usgs.gov/nmpstds/dlgstds.html. Accessed August 30, 2010
22. US Geological Survey (1999) http://rockyweb.cr.usgs.gov/nmpstds/nhdstds.html. Accessed August 30, 2010
23. Spatial Data Transfer Standard Technical Review Board (1997) Spatial Data Transfer Standard (SDTS) – Part 2, Spatial Features, Draft for Review. Federal Geographic Data Committee. http://mcmcweb.er.usgs.gov/sdts/SDTS_standard_nov97/p2start.html. Accessed August 30, 2010
24. US Board on Geographic Names (2010b) Geographic Names Information System (GNIS). US Geological Survey. http://geonames.usgs.gov/pls/gnispublic/f?p=139:8:736061011105747. Accessed August 30, 2010
25. Lakoff, G (1987) Women, fire, and dangerous things. University of Chicago Press, Chicago
26. Usery EL (1993) Category Theory and the Structure of Features in Geographic Information Systems. Cartography and Geographic Information Science 20 (1): 5–12
27. Mark DM, Turk AG (2003) Landscape categories in Yindjibarndi: Ontology, environment, and language. In: Kuhn W, Worboys M, Timpf S (eds) Spatial information theory: A theoretical basis for GIS, lecture notes in computer sciences 2855, Springer-Verlag, Berlin
28. Kavouras M, Kokla M (2008) Theories of Geographic Concepts, Ontological Approaches to Semantic Integration. CRC Press, Boca Raton, Florida
29. GeoNames (2010) GeoNames Ontology. Geonames.org. http://www.geonames.org/ontology/. Accessed April 30, 2010
30. Dolbear C, Hart G, Kovacs K, Goodwin J, Zhou S (2007) The Rabbit Language: description, syntax and conversion to OWL. Ordnance Survey Research. http://www.ordnancesurvey.co.uk/oswebsite/partnerships/research/publications/docs/2007/Rabbit_Language_v1.pdf. Accessed March 19, 2010
31. OpenCyc (2010) OpenCyc.org. http://sw.opencyc.org. Accessed February 8, 2010
32. Herring JR (ed) (2006) Open GIS implementation specification for geographic information – Simple feature access – Part 1: Common architecture. Open Geospatial Consortium Inc., OGC 06–103r3, Wayland, Mass
33. Caro H (2010) Analysis of Spatial Relation Predicates in US Geological Survey Feature Definitions. US Geological Survey Open File Report, US Geological Survey, Reston, Virginia
34. Allemang D, Hendler J (2008) Semantic Web for the working ontologist, effective modeling in RDFS and OWL. Morgan Kaufmann, Burlington, Mass
35. Guarino N (1998) Formal ontology and information systems. In: Guarino N (ed) Proc. 1st International Conference on Formal Ontologies in Information Systems (FOIS'98), Trento, Italy. IOS Press, Amsterdam
36. Varanka D (2009) A topographic feature taxonomy for a U.S. national topographic mapping ontology. In: Proceedings of the International Cartography Conference, Santiago, Chile. http://icaci.org/documents/ICC_proceedings/ICC2009/html/nonref/9_7.pdf. Accessed August 30, 2010
37. Varanka D, Jerris T (2010) Ontology Patterns for *The National Map*: Proceedings AutoCarto 2010-ISPRS Commission IV – ASPRS Fall Specialty Conference. Orlando, Florida
38. OntologyDesignPatterns.org (2010) Semantic Web portal, NeOn Project. http://ontologydesignpatterns.org/wiki/Main_Page. Accessed August 30, 2010
39. Sheth A, Ramakrishnan C (2003) Semantic (Web) Technology In Action: Ontology Driven Information Systems for Search, Integration and Analysis. In: Dayal U, Kuno H, Wilkinson, K (eds) IEEE Data Engineering Bulletin, Special issue on Making the Semantic Web Real, IEEE Technical Committee on Data Engineering

40. Torres M, Quintero R, Moreno M, Fonseca F (2005) Ontology driven description of spatial data for their semantic processing. In: Rodriguez MA, Cruz IF, Egenhofer MJ, Levashkin S (eds) GeoSpatial semantics, first international conference, Mexico City, Mexico, Proceedings. Lecture Notes in Computer Science 3799, Springer, Berlin, Germany
41. Keßler C, Janowicz K, Bishr M (2009) An agenda for the next generation gazetteer: Geographic Information Contribution and Retrieval: ACM GIS'09, November 4–6, 2009, Seattle, Washington
42. Herskovits A (1986) Language and Spatial Cognition: an interdisciplinary study of the prepositions in English. Cambridge University Press, Cambridge

Index

N. Ashish and A.P. Sheth (eds.), *Geospatial Semantics and the Semantic Web: Foundations, Algorithms, and Applications*, Semantic Web and Beyond 12, DOI 10.1007/978-1-4419-9446-2, © Springer Science+Business Media, LLC 2011